To My Long Time Friend Barbara Pace —

Best Wishes and Happy Landings

Always

Bill Pohlmann

Sunday Pilot

Airborne Tales of a Weekend Flyer

Sunday Pilot

Airborne Tales of a Weekend Flyer

by Bill Pohlmann

Printed in the United States of America

Cover by Adam McCaulley

ISBN: 1-891231-43-X
Library of Congress Control Number: 2001088894

Word Association Publishers
295 Fifth Avenue
Tarentum, Pennsylvania 15084
800-827-7903
www.wordassociation.com

CONTENTS

DEDICATION

This book is dedicated to everyone interested in becoming a co-pilot to adventure by discovering what it feels like to fly a single engine, light airplane. This 'everyone' includes those readers who are currently flyers, or future flyers, or armchair flyers.

For those of you who are just growing up and are still too young to take part in the world of sport aviation, at least this book will help you to know what it feels like to fly a small airplane and to see what a pilot sees and to go where a pilot goes, in good weather and bad.

This book is also dedicated to all those pilots who are not lucky enough to live where there is no winter. This book will help you to pass the time on long winter evenings, by re-living the fun, thrills, and dangers of flying small airplanes while patiently waiting for the coming of spring.

Bill Pohlmann
Palos Verdes,
California

FOREWORD

This is a book for those who do not fly but always wanted to know how it feels to sit above the Earth and manipulate the controls of a single-engine airplane as it drones across the endless sky. *Sunday Pilot, Airborne Adventures of a Weekend Flyer*, by Bill Pohlmann, puts the reader into the cockpit next to the pilot and allows him to experience what the pilot feels and sees as they cruise a mile above the Earth where there are no roads or sign posts to lead the way.

This book portrays flying adventures in all kinds of weather. Even the armchair pilot who never leaves the ground — but always wanted to be a co-pilot to adventure — will find plenty of emotion-charged adventure encountered by his airborne brethren. Bill Pohlmann has captured the full range of emotions affecting pilots — from the anxiety of being stalked by thunderstorms to the sun's changing brightness as weather conditions change, from bright to dim, psychologically reducing a student pilot's confidence during a long solo, cross-country flight over the barren and vast Arizona desert.

Most flying books describe how to obtain a pilot's license, or once licensed, how a pilot can improve his skills. But Pohlmann's book is different. It focuses on how it feels to fly small airplanes under a variety of conditions. The book also discusses many of general aviation's activities such as social fly-ins hosted by 'type clubs,' pursuing a variety of challenges, using the family airplane as a recreational vehicle, and hanger-flying.

Besides the airborne adventures, the many colorful people that Pohlmann describes during his twenty-five years of 'messing around' with airplanes is reason enough to read his book. It is well written and a good and easy read, especially Chapter Eight: Hangar Flying Without a Parachute, Airborne Adventures at Ground Level. The people described in this chapter are those to whom everyone can relate.

This book should appeal to both pilots and nonpilots. For those who fly, this book describes situations to which you will readily relate and bring back fond and precious memories.

Capt. Barry Schiff, TWA (retired)
Aviation Writer and Safety Consultant
Los Angeles, California

Chapter One
Sunday Pilot

The What, Who and Where

WANTED
CO-PILOTS FOR ADVENTURE
To Fly Single Engine Airplanes In The Yukon
NO EXPERIENCE NECESSARY

*** * APPLY INSIDE * ***

What would you do if you saw a sign like the one above? I know what I'd do. Since I was twelve, I've dreamed of being a Bush Pilot in Alaska. And, if I ever saw a sign like that, I'd run inside and sign up before their quota was filled.

Have you ever heard the droning of a small airplane as it crawls slowly across the cloudless sky? Did you ever watch it wondering what kind of a person flies something that small? And, as you watched it, did you ever wonder where it came from and where was it was going?

Everyone at one time or another has been curious about small airplanes, especially if it's a hot day and you'd rather be someplace else. As an airplane moves across the sky, you might even wish that you were in it and that it was taking you away. Or, maybe you're twelve years old, and as you watch a small airplane fly overhead, you wonder what it feels like to hold the controls in your hands and to make the airplane maneuver; or what it sounds like, sitting in the cabin, just behind the engine? Or, what the world looks like from a half a mile up. You might even wonder how a pilot knows which way to go without roads to guide him.

This book is not about what it takes to be a pilot of a small airplane, or how I got my Private Pilots License. It's about my lifetime love affair with small airplanes. It's about what it feels like to fly them, and about the adventures I've had while flying and about the people I've met and flown with along the way, and the flying adventures they've had.

This book will help you to feel what a pilot feels as he flies his single engine airplane. You'll feel his fear in an emergency or when he's about to make a mistake that may cost him his life. And you'll also feel the fear a student pilot feels as he arrives at the end of a long solo cross country flight, only to discover that he's not at his destination and is lost and low on fuel. Sit next to a student pilot in the right seat, and listen to his 'internal dialogue,' as he talks to himself to impose discipline and keep panic from controlling his actions.

Fly with a high time pilot with thousands of flight hours who loses control and judgment and makes an almost fatal mistake. Or, sit next to a Sunday Pilot with fifteen log books of documented Pilot-In-Command time, as he flies a low performance airplane into a thunderstorm. Feel what he feels, as the panic wells up inside him and the fear of death turns into a sharp pain in his gut. Or, be in the cockpit of an Antique tube and fabric airplane while it tries to fly over the Rockies, when a mechanical problem occurs that turns a routine flight into one of high adventure.

All through my youth I looked for a book that would tell me what it feels like to fly small airplanes—to understand what it's like in the sky in the pilot's world where there are no roads. And, since I never found a book like that, I took the time to sit down and write one.

•

Is a small airplane easier to fly than an airliner? Can anybody fly airplanes? Who are these people that feel safe flying a small airplane across the gigantic sky?

What Is a Sunday Pilot?

Sunday Pilots are people just like you, young or old, man or woman whose love of airplanes has led them to take flying lessons and to receive their Private Pilot's License. And, they enjoy flying so much, they make it their hobby and weekend recreation. Some of them may own their own airplanes, or if not, they rent airplanes from their local flight school. They're rarely commercial pilots, they have non-flying 'ground' jobs' during the week. And, when they fly on the week ends, it's just for recreation and the trips they go on are for pleasure, not work.

The term 'Sunday Pilot' comes from the related 'Sunday Driver' to which it's very much akin. Sunday Drivers do not drive regularly. And, because of their infrequent use of the highways, their driving habits are not always predictable to the more experienced general driving public.

Commercial pilots consider those of us of 'general aviation' who don't fly for a living, as unskilled and they lump us into the same category with student pilots. Now, student pilots are regarded by everybody in aviation as a menace, unskilled, to be avoided at all costs. Student pilots are considered dangerous, because they too are unpredictable.

I'm not sure that Sunday Pilots are quite as unpredictable as their 'professional brothers' would have the public believe. But, I admit that our skill level is not as high as those who fly for a living; usually in high performance equipment that requires advanced training and a higher skill level. The 'pros' don't fly for fun. Their flying is serious and a world apart from ours.

The Sunday Pilot's Attitude

Weekend flyers come in all sizes, shapes and attitudes. Some approach flying seriously and often act aloof, as if piloting skills were handed down from God, and a skill capable of being mastered only by those of superior intellect. They swagger around the airport like the 'ace of the base,' even if their log book still documents their flying time in hundreds of hours rather than in the thousands.

By contrast, others are cavalier and unconcerned about the inherent risks of flight. They act as if flying was as simple as riding a bicycle, as if safety measures such as weather briefings, preflight inspections and checklists are just a bother.

These two extremes represent both ends of the 'mentality' spectrum of the Sunday pilot. It should be pointed out, however, that most weekend flyers fall somewhere in between, considering flying as their hobby and a labor of love. To them, it should always be safe and above all, fun.

The Airspace

I think most Sunday Pilots try to stay out of the way of the professionals in their 'heavy iron.' And for the most part, our equipment is small and of low performance and usually very unsophisticated. And, the majority of our airplanes are single engine which dictates pretty much where and how we fly: low and slow, usually below ten thousand feet, and usually below 150 mph.

Not only is the commercial equipment multi-engine, very large and expensive, but it's also designed for high performance and passenger safety. Commercial airplanes used for hauling passengers normally contain dual or triple redundant systems, computerized and capable of flying and navigating the airplane automatically.

And, for most of each flight, the pilot doesn't even have to touch the controls if he doesn't want to. He can just sit there managing things.

So at the outset, it should be understood that there is really no conflict between general aviation flyers and commercial pilots, whether corporate or airline. The fact is, most of the time us weekenders don't even share the same part of the sky. Airliners flying under positive radar control, on instrument flight plans, normally cruise the friendly skies above twenty four thousand feet.

For corporate and airline feeder service, that's still flying pressurized propeller airplanes, like fifteen passenger twins. Their altitudes are usually between fifteen thousand to the low twenties. Us Sunday flyers in our single engine, 'Belch Fire 8 Special,' with its tiny one hundred and fifty horsepower, four cylinder engine, normally never get any higher than six thousand feet, except when flying over local mountains. So, the notion that there's a fight for the sky just isn't true. It's really not that crowded.

What usually keeps a general aviation pilot low and slow is his pocket book. For the airplane renter, the fees are just too steep for higher performance equipment, and not really worth the extra cost. On an average trip of two hundred miles, the difference in travel time between a slow airplane and a faster one is only a little more than a half an hour. So, if a half hour is worth twice the rental cost, then enjoy! But, if it's a 'life and death' emergency to get somewhere fast, then you are better off flying the friendly skies of United!

From the perspective of the Sunday pilot, flying is fun, and speed is just a number. And, 'getting there' isn't just half the fun, it's all the fun! And for those of us that own our own airplanes and fly for pleasure, flying high and fast isn't a priority and isn't where the fun's at. The real fun of flying is low and slow, where one can look down and see the countryside drifting beneath the wheels without fighting the stream of endless cars on the highway, Besides, even on a trip when flying as slow as one hundred miles per hour, getting there is still three times faster in an airplane, than driving.

Once a single engine airplane gets above ten thousand feet there's not much to look at but haze. And all that can be done at that altitude is to just sit back and 'crank off the miles.' Not exactly why week end flyer's 'flap their wings.'

The 'Flight Time' Paradox

To me, one of aviation's most curious practices is the way a pilot's flight experience is measured. The Feds seem to think that the flight hours recorded in the Pilot's log book are the best estimate of his skill and judgment. Therefore, a commercial pilot who flies about one hundred hours per month, or a thousand hours each year, has a much higher experience and skill level than those of us who only fly now and then for fun.

That's probably an accurate measurement of a commercial pilot's ability; but, I think that using that method to measure the skill level of Sunday Pilots becomes a paradox. Weekend flyer's may or may not fly often and the yardstick of Pilot-In-Command (PIC) time doesn't predict the same level of experience when applied to 'ussuns.'

For most of us private pilots who log only a couple hours each weekend, the annual total is between 40 and 100 hours per year. That's less than what a commercial pilot flies in a single month. This difference is important, because it means that weekend pilots are flying ten times less often than Commercial Pilots, and in flying the frequency is almost more important than one's total flight time. It's a matter of a pilots reaction time, or how fast a pilot can recognize a problem and take appropriate measures. The less one flies, the longer it takes to react. The longer one's reaction time, the less safe is the airplane. Pilot's call this 'rhythm.' And when they haven't flown for a while, they talk about getting their 'rhythm' back.

So, after twenty five years of observing Sunday Pilots of various skill and experience levels, I think that a flyer's skill measured by using his total PIC time is really unrealistic and a paradox. I think it should only be used when determining the experience level of commercial pilots, not private pilots. This paradox is: 'the more PIC hours a pilot has in his log book, the safer and more competent he is as a pilot.'

This measurand of hours equating to skill level becomes a paradox when applied to Private Pilots who only fly infrequently. A Private Pilot may have an abundance of flight time in his log book, but because of his individual insensitivity to danger or his cavalier attitude, his PIC time gives an evaluator a false impression of his flying competence. I've known some 'high time' Private Pilots that have no respect for the dangers of flight because of their extensive experience. Their experience seems to corrupt their judgment..

After watching 'high time' and 'low time' Sunday fliers, I maintain that competency really depends on the individual Pilot. And my flying experience leads me to believe that flight experience alone doesn't equate to flight judgment or a higher level of safety. Some people fly safer with better judgment the more they fly; while others become de-sensitized to danger, becoming complacent with additional flight time. Once a pilot is not impressed by danger, or looks forward to it like an 'adrenaline junkie,' his chances of dying in an airplane go way up!

To illustrate these two extremes of the paradox, I offer the examples Bob Johnson and Tom Stansbury. Bob Johnson illustrates the low-time pilot who gets into trouble because of his lack of experience, while having very low numbers in his flight log, and shows just how quickly a routine flight maneuver can turn into one that's'life threatening.' Tom Stansbury illustrates the pilot with over fifteen thousand flight hours and is lulled into a false sense of security because of his years of experience. His judgment is impaired to the point of inaction, until it is almost too late!

•

Bob Johnson

In the summer of 1948, Bob Johnson was a portly man in his late fifties, medium height, thick black hair, wire rimmed glasses, an infectious smile. He always smiled when he talked. And, because he always smiled, I considered him to be a truly happy man. And since I was only twelve at the time, I always called him 'Mister' Johnson.

Mister Johnson owned a cottage on Bass Lake, a small private lake forty miles east of Cleveland, Ohio, near the farming community of Chardon. Chardon was famous in those parts for its annual Maple Syrup Festival. Mister Johnson's cottage was four lots down from my Grandfather's, where I spent my summer weekends.

Bass Lake was nestled in the middle of lush farmland and surrounded by thick woods, about a half hour's drive from Chardon. The Lake was the summer home to fifty families; forty three on the Lake's west shore that were clustered in an eight block area, a quarter of a mile from the water; while the remaining seven were spread along the water's edge on the Lake's eastern shore.

The residents of the lake's west side were isolated from those of the east, as there were no connecting roads around the lake. Private farmlands abutted the lake's perimeter, except for it is shoreline on part of the west and east sides. To get from one side of the lake to the other meant driving into Chardon on two different county roads, and then back tracking out of Chardon on a different road that passed the west side on the Lake. Driving over tertiary country roads from one side to the other, took almost an hour.

Of the seven cottages on the east shore, my Grandfather's was located in a woods near a large creek that emptied into the lake; while the Johnson's place was fifty yards away in the middle of a meadow. The wonderful thing about the seven cottages was that each of them had waterfront property with a boat dock.

During the summer the main attraction at the lake was bass fishing, and the fishing was good. And in the summer of 1948, the war had been over for a year and the peacetime economy was beginning to flourish again. And that was also the summer that Mister Johnson gave me my first airplane ride.

Bob Johnson was the only pilot anyone in our family knew personally. And just after the war when general aviation was allowed to come back to life, Bob had bought a new airplane and learned to fly. His airplane was a yellow and red Aeronca Champion, a high wing monoplane, constructed of steel tubing and Irish linen, with two coats of silver non-tauntening Butyrate dope, and twenty coats of yellow and red lacquer.

Bob wasn't a pilot in the war, since he was in his fifties when the war broke out, but he was a man of means, and had always wanted to be a flyer. He was an Optometrist by trade, owning his own practice. And by 1946, when civilian airports began coming to life, flying lessons were offered free by flight schools everywhere to anyone who purchased a new airplane. And since Bob had dreamed of someday being a pilot, he took advantage of the opportunity and bought a new airplane. Within a few months, he'd graduated with his Private Pilot License.

Bob kept his airplane at a small country airfield just two miles as the crow flies, to the east of Bass Lake. But after a summer of weekend flying at his cottage, he got the idea that it would be twice as convenient if he were to base his airplane at the new seaplane base on Lake Erie, near his house in Cleveland. Then he could commute by air in the summertime to his cottage at Bass Lake. This, he reasoned, would not only save him the two and a half hour drive to the lake, but it would be a lot more fun flying over the traffic than driving in it. To do this, however, meant changing the airplane's landing gear from wheels to pontoons, as floats were called in those days, but that didn't seem like a big deal to Bob.

The seaplane base had been built at the edge of Lake Erie the year before. It was constructed at the water's edge at Cleveland's lake front, near the Cleveland Indian's baseball stadium. The idea seemed to make more and more sense to Bob, the more he thought about it. And through the winter months, Bob thought of nothing else, as he spent happy hours, dreaming of how it would feel the next summer as he would fly his own airplane on pontoons from Lake Erie to Bass Lake. Taking off after work on a Friday afternoon and flying directly to his cottage, where he could tie up at his own boat dock. No traffic, no hassle.

He reasoned that flying to and from Bass Lake would save him the long drive that he and his wife had endured for years. Of course, they couldn't bring along much baggage in the airplane, as it only had two seats, one behind the other, and only had room for ten pounds of baggage when both seats were occupied. But, he thought they could always leave one of their cars at the lake for trips into Chardon for groceries.

The more Bob thought about it, the more intriguing the idea became. So, by spring, he took lessons at Cleveland's seaplane base on how to land an airplane on the water, and by that summer he began flying his airplane to and from Bass Lake.

•

Float plane flying has a certain inherent risk that flying land planes don't have. Every time a water landing is attempted, there is a risk that unless the pilot is careful and really concentrates, the chance for disaster is very real.

All went well for Bob's water flying at Bass Lake that summer, as he landed his seaplane every Friday evening just before sunset, and took off for home every Sunday evening at dinner time. The distance was only sixty air miles and usually took about an hour's flying time. The pontoons slowed the airplane's cruise speed from eighty five miles per hour to about seventy five, but when it comes to seaplanes, speed doesn't matter, since it's the utility that's important. And to Bob and his wife, because it was such a pleasure flying at a thousand feet over the traffic passing all of the cars, air speed never mattered. Whatever time it took was worth it when they landed at the lake and water taxied up to their boat dock in front of their cottage. The cottage door was only steps away from the dock and made their trip each weekend a relaxing pleasure.

Bob's routine of taking off and landing at Bass Lake worked well all summer. He'd begun in early June and now in mid-August, the routine was becoming almost second nature, not to mention the fact that he now had twenty hours of water flying Pilot-in-Command' time neatly documented in his Pilot's log book, plus the five hours of lessons he'd taken at the seaplane base to get his Seaplane' rating. And after a summer of water flying experience he was feeling pretty cocky about his ability as a seaplane pilot. But, as the August temperatures and humidity climbed into the nineties and stayed there day after day, Bob's luck was about to change.

Seasoned pilots can easily predict the dangers of hot weather flying. But to a pilot who considers himself 'experienced' with only a couple dozen hours of water flying, Bob's attitude concerning margins of safety became cavalier. Even for a beginner, it's strange that he never accepted the fact that in very hot and humid weather, airplane performance drops dramatically, especially during the hottest part of the summer.

Now, Bass Lake is a small lake, as mid western lakes go, with not nearly the room required for seaplane operations. It's kidney shaped with a peninsula jutting so far out into its center from the eastern shore, that if one was fishing below the point of the peninsula at either end of the lake, the other end of the lake was obscured from view. The lake was only a quarter of a mile wide where the peninsula protruded, then widened towards both ends; while its length was just three quarters of a mile.

And since it was a fairly small lake, it was also shallow; only twenty feet deep at its center. And since it was shallow, thick weed beds grew outward from its shore for hundreds of feet, reducing the open water area. The weed beds grew so thick that a barge with a weed cutting bar mounted on it spent several weeks at both the beginning and the middle of summer cutting the weeds to provide more 'open' water for fishing.

As with most lakes in the midwest, Bass Lake had no beaches, and was mostly surrounded by woods and farmlands, with the tree line growing right down into the water. There really was no edge of the lake that was solid ground. And although it wasn't very large, it was still the private weekend haven for 50 families of hard core fishermen. And since it was not a public lake, residents cherished their solitude and seclusion from the cares of the world.

In the heat of the summer, when the air is as still and stifling and humid as a jungle, the air turns into visible moisture in the form of a white haze, not unlike fog. And, since the air is totally saturated with moisture, it offers very little in the way of lift for a small, under powered seaplane. In fact, to operate a seaplane under such conditions would be to incur great risk.

Now, flying a seaplane in the heat of a summer is a lot like mountain flying; the sun's heat affects both. In the mountains, the rule is to fly before ten o'clock in the morning or after five o'clock in the afternoon; but never during the hottest part of the day. The air just gets to too thin to offer any lift to an airplane's wing.

Whether Bob ever knew about such hot weather Bush Pilot 'tips' is not clear, but to a flyer, the distance required for a seaplane's takeoff run on a hot day at a lake as small as Bass Lake could predict that it would pretty much take up the entire length of the lake.

And, even once airborne, the Lake's tree line was an additional problem. For a seaplane to be able to clear the trees at the water's edge, it would have to climb to an altitude of at least thirty feet. But to clear those trees after a water takeoff on a sultry day was asking an awful lot from an Aeronca Champion, with it's modest sixty five horsepower engine. The land plane version of the Champion was always thought of as being totally under-powered; making the seaplane version even worse. The pontoons themselves not only weighed more than the land plane's wheels, but once airborne, offered nothing but drag.

The fact that Bob had been able to fly in and out of the lake all summer without a mishap was only a coincidence. He just happened to arrive or depart during the cooler part of the day, during late afternoon or early evening. And, once at the lake during the day, he never flew the seaplane. Additionally, the most important thing that kept Bob out of trouble was the fact that Bob had an awful lot of luck going for him. But, as fate would have it, his luck was about change one Saturday in mid-August.

At Saturday noon Bob received an urgent phone call from Cleveland, concerning a personal business matter. His presence was required at an important meeting that should only last an hour. He thought he could fly there and return to the lake before the end of the afternoon. Since it was imperative that he attend he thought he could expedite things by leaving his wife at their cottage, and could turn the whole thing around within three and a half hours.

•

For most fishermen, the middle of the day was always too hot for the fish to be biting, fish go deep where the water is cooler, only feeding close to the surface during the early morning and early evening. But, despite common sense, at noon, three hardy fishermen were on the water to try out their new fishing tackle.

The three had just arrived at the lake an hour before for the weekend. And each had just purchased the new state-of-the-art Shakespeare Spinning rig, the top of the line of fishing tackle. In 1946, spinning rigs were the newest rage and were fast replacing fly fishing as the rig of choice. The spinning rig was a cross between a bait and fly casting reel; it looked like a bait casting reel, but acted like a fly casting reel. And the three fishermen wanted to be the very first on the lake with the 'newest and the best,' not to mention the most expensive.

The fact that the fishermen were out in the stifling heat, in itself, was not of much concern, but the fact that they chose to row over to the southern end of the lake, below the point of the peninsula was important. They rowed to the edge of the weed beds that grew outwards from the shore towards the open water for more than fifty yards. During early evening, those same weed beds were a favorite feeding spot for Large Mouth Bass. So the three decided that if any tackle could catch fish during the worst part of the day, their new Shakespeare rigs were the ones.

At the northern end of the lake, beyond the peninsula, the growling sound of an airplane engine coming to life could be heard. The three fishermen, at the other end of the lake, heard the initial coughing of an engine, but after a few seconds when the sound subsided they dismissed the thought that Bob and his airplane were about to go flying. In the blanket of heat haze that lay like fog a foot above the water, sound doesn't travel very well.

Once the engine caught, Bob immediately retarded the throttle, as he began to slowly water taxi to the northern end of the lake.

As he approached the thick weed beds that grew some distance from the shore, he turned the plane around so that it pointed south, where his takeoff run could use the full length of the lake, as modest as that was.

With the oppressive temperature and the haze most of the engine noise was attenuated, as the seaplane began its takeoff run. On a cooler day, sounds of the engine at full power could be heard reverberating for several miles; but, on this hot, August noon, sound was greatly diminished.

With the throttle wide open, the airplane began moving very slowly at first, and then after what seemed a long time, it finally began picking up speed, as the pontoons sluggishly plowed through the glassy water, like two lethargic torpedoes.

A dozen seconds later, the fishermen began to hear the angry snarl of the Aeronca's engine at takeoff power coming from the other end of the lake. As they sat still in their boat, eyes straining into the haze, it was impossible to tell the seaplane's direction. And, because the peninsula jutted into the Lake's center, the view of the other half of the lake was obscured.

As the engine's sound increased, the fisherman rowing the boat began moving it away from the on coming sound. He tried repositioning the boat from what he thought was the airplane's path. And, as he was rowing the two others squinted into the bright haze, hand's cupped to their eyes hoping for a glimpse of the airplane.

At the controls, Bob pushed the airplane's nose down, forcing the pontoons to come up 'on step' so they would begin to 'plane' and pick up flying speed. As hard as he tried to hurry the takeoff process, the longer it seemed to take. The airspeed was not increasing as it normally did. and Bob was suddenly surprised to discover that his airplane wasn't performing the way he expected it to.

He glanced at the point of the peninsula as it passed by. Half the lake was already behind him, and the pontoons were just now barely beginning to plane.

Panicked, he jerked the control stick back, raising the seaplane's nose and horsing the pontoons off the water and into the air. But, although the seaplane was airborne, it still hadn't reached flying speed; it was just mushing along a foot above the water in 'ground effect.' It had to gain flying speed to be able to climb, and couldn't remain in ground effect for very much longer, as the seaplane was fast running out of lake.

Gently, Bob eased the control stick forward, lowering the seaplane's nose to the horizon and allowing it to pick up flying speed. He looked ahead. The trees of the southern shore were growing larger in the windshield. What should he do? Abort the takeoff?

He looked at the airspeed needle. It was pointing to sixty two miles per hour, too fast really to abort. He'd already run out of lake. He'd waited too long.

Through the windshield the view of the southern shore was rushing at him. He was going too fast to stop, and even if he could get the seaplane back down on the water, the trees were too dense and their trunks too thick for a pilot to survive a head-on crash at sixty two miles per hour in a fabric covered airplane.

When the three fisherman caught a glimpse of the seaplane as it came out of the haze passing the point of the peninsula, their boat was already a hundred feet off to one side of Bob's takeoff path. They sat transfixed, as Bob's sea plane continued onward, its pontoons now several feet above the water.

Bob looked up and saw that the lake was about to vanish beneath him. He quickly looked around for a break in the woods, or where the trees were smaller. He noticed that there was a somewhat barren spot in the tree line just thirty degrees to his left. The spot was about ten feet wide, where no trees had grown. It was not really wide enough for his entire seaplane, but wide enough for his pontoons to clear without snagging anything.

Without hesitation he altered his course and carefully banked the seaplane to the left, aiming for the gap in the tree line. If everything went well the pontoons might just clear–even if the wings clipped the thin branches of the tops of the other trees, there was a good chance he'd survive.

With the seaplane floundering on the ragged edge of a stall, he nursed it's nose several degrees higher. The Aeronca reacted sluggishly in the steamy air, reluctantly climbing out of ground effect.

But in his haste to clear the trees, Bob hadn't noticed the three fishermen in their boat, now directly in his path. And as he maneuvered the seaplane, the fishermen suddenly realizing that Bob was now heading directly at them, began shouting and waving for him to turn away. The man at the oars began rowing with all of his might, splashing and chopping at the water with short strokes, trying to move the boat and escape what he believed to be his imminent death.

Bob's eyes were attracted to the sight of the boat by the movement of the fishermen's frantic waving — their final signal to him to alter his course and save their lives. But the time was gone, and the seaplane was closing with the tree line at the rate of one hundred and two feet per second, and the shore line was now less than two hundred feet away. In a final desperate effort, Bob raised the airplane's nose even higher and entered the beginnings of a stall.

The entire airframe shuddered and the engine's growling sounds at full power changed to a deafening wail. Helpless and paralyzed with fear, Bob sat motionless, now just a passenger in his own seaplane–no longer in control. He'd made his 'life and death' decision and nothing could change it.

Seconds later, the last fifty yards between the airplane and the boat evaporated, as the seaplane roared by the boat with just feet to spare, it's pontoons slicing through the air just inches above the fishermens' heads. The fishermen, fearing for their lives and deafened by the howling blades of the seaplane's propeller, dove straight into the lake at the very last instant, overturning their boat and losing forever their new Shakespeare spinning tackle.

•

Needless to say, after that incident, the Homeowner's Association banned forever Bob's seaplane from the lake. Some of the Association thought that Bob and his wife should sell their cottage and leave the lake for good, as the residents were certain that Bob's seaplane was a menace and kept the fish from biting. They reasoned that the noise of the pontoons slicing through the water at sixty miles per hour had to scare the fish. They were certain that if the seaplane was allowed to continue using the lake, the good fishing that they'd enjoyed for years would be gone.

Bob's airplane was never seen at the lake again, at least not at ground level. The pontoons were taken off and the wheels put back on and the airplane was returned to its former base, the airfield two miles from the lake. And from that time on Bob and his wife resumed the two and a half hour drive from Cleveland to their cottage. And the Aeronca Champion flew only on summer weekends when Bob was at his cottage and tired of fishing.

It was shortly after Bob had his airplane's wheels reattached that he gave me my first airplane ride. And during the ride, for some reason he felt compelled to fly low over Bass Lake, buzzing the fishermen who dotted its surface in their row boats. I guess he figured that if he couldn't fly off the lake anymore, at least he could buzz it now and then.

•

Tom Stansbury

Tom Stansbury is the kind of guy every man wishes he could be. He had only one flaw, five ex-wives! Tom always reminded me of 'Terry', the 'Terry and the Pirates' character; a cross between a soldier of fortune and 'Indiana Jones.'

Tom is an original. For one thing, he is the original 'marrying fool!' When I first met him, he was in his late forties and tall with black hair, the kind of hair women love to run their fingers through. He had an athletic build and a deep resonant voice that gave one the impression of total sincerity, which was a large part of his charm. When he spoke, his voice filled the entire room and its tone was very confident and reminiscent of Henry Fonda. It was a voice that a woman could lose herself in. And when he smiled, which was often, his teeth were perfect and white. The total effect of his smile, voice, and manner was just too much for women to resist.

He told me one time, while we were flying along the coast of southern California in his very large, all metal high winged, five passenger, round engined Cessna 195, that if he ever had it to do over, he'd cut off his 'privates' with a meat cleaver, thus saving himself the cost of all of the alimony payments and 'woman' problems he'd had over the years. It seems, that if he'd done something close to that years ago, he'd have saved himself from five wives and would still have a very large bank account.

•

In his time, Tom had made a mountain of money. And when I first met him, in his spare time on weekends, he was building two retirement homes in Baja; one on the inland side of the peninsula near San Felipe, on the Sea of Cortez — which was his winter house; and the other south of Ensenada, on the western shore of the peninsula, his summer house. And every Friday afternoon he would get into his airplane after work and fly to Mexico to continue building his houses.

Those houses were supposed to be for his retirement, but after the cost of each divorce and alimony payments he was forced to liquidate them, along with his dreams.

One time he was flying down to Baja to work on his houses, in his Piper TriPacer, the tube and fabric, high winged, tricycle landing geared airplane, of the early fifties. It was the airplane he'd bought after he sold his Cessna 195. As he crossed Mission Bay in the San Diego area, the TriPacer's engine started coughing and then quit, all within ten seconds. It was summertime, and as he looked down at the beaches of Mission Bay there were sunbathers lying on every possible patch of sand that could be used for an emergency landing. He wanted to land on the beach, but there was no way he could do it with all those people. So he ditched the airplane in the bay, fifty feet from shore, in five feet of water.

Now normally, ditching an airplane is very tricky. When it hits the water with fixed landing gear, it instantly noses over, upside down. But Tom was 'blessed.' He was a very competent pilot with fifteen thousand flight hours documented in his log books, having flown for almost thirty years by that time, and was able to ditch the airplane in calm water and was able to land flat so that the airplane didn't invert or sink immediately, In fact, he got a group of nearby swimmers to help him pull the airplane to the beach before it could sink.

The FAA Accident Investigator told him later that the

engine stoppage was due to a certain strain of Bacteria that thrives on gasoline and grows into large colonies of algae-looking slime. And it was bacteria contamination that blocked his fuel line and killed his engine. The entire fuel system of the TriPacer was clogged with the stuff. The investigator thought that it was Tom's constant use of contaminated fuel from Mexico that caused the problem.

•

Over the years, Tom had owned a string of airplanes. When he first took me flying it was in his current airplane, a grand old classic of the fifties, a five place Cessna 195, with its giant round 300 horsepower engine of large displacement. The 195 was a very large all metal, single engine airplane built by Cessna in 1948. At the time it was used as a corporate airplane. And, from the ground, a Cessna 195 could be heard, like a helicopter, five minutes before it could ever be seen.

Before the 195, Tom owned a Beechcraft Bonanza, and before that a Piper TriPacer, and before that a Cessna 170 tail dragger. He alternately sold and bought back the TriPacer every few years; and by the time I'd first met him, he'd already owned it and bought it back four times. I guess Tom had a hard time making up his mind about women and airplanes.

During the late eighties, after not seeing Tom for ten years, I ran into him one day at a fly-in at the Camarillo Airport. He looked a little older, but not much. He still looked like someone a lady would want to go home with. But, for once he had a problem that was worse than five ex-wives: Cancer. After several surgeries, the doctor made him give up drinking and smoking. He said he hadn't had any 'fire water' in almost eight years. And, as we caught up on old times, could see that though his hair was gray, he was still the same old charmer.

He was still flying the TriPacer, although it looked awful–its headliner was removed and exposed tuffs of insulation between the bare control cables. Some of the insulation was hanging out and looked like the scarecrow from the 'Wizard of Oz.' The instrument panel had a lot of vacant holes where there should have been important and mandatory flight and navigational instruments. On the whole, the airplane didn't appear to even be airworthy. But, the most noticeable missing item from the panel was the whiskey compass.

Sometimes, Tom reminded me of 'Sky King', the consummate pilot, because of his resourcefulness and self-reliance in a tight situation. When you traveled with Tom, things happened to you that wouldn't happen to you if you were traveling alone, or with anyone else. What usually happened to Tom was women. Women just wouldn't leave Tom alone. They flocked to him like iron filings to a magnet. Even when I was with him, women would 'come on' to him. He never encouraged them; they just felt compelled.

Even before I ever took flying lessons and knew one end of an airplane from the other, Tom would take me along when he went flying, just for the company. I guess I was a good audience for him and, since I was crazy about airplanes, it was always a thrill for me to go for an airplane ride.

When you flew with Tom, he always assumed that, no matter who you were, you knew as much about flying as he did. So one summer afternoon, I found myself in the plush leather right seat of his Cessna 195, at three thousand feet over Laguna Beach, flying northward back to Long Beach. We'd taken off from Long Beach earlier for Palomar Airport, as Tom was having problems with one of the airplane's radios and needed to have it fixed by the Radio Repair shop at Palomar. And, as we were winging our way back, I was gazing out the copilot's side window watching the landscape move slowly by, when suddenly, Tom turned to me and said, 'Take It!'

'Take what?' I asked, looking over at him.

'Take the controls,' he said, motioning to the control wheel in front of me with his head. 'I gotta look somethin' up.'

And with that, he let go of the wheel and pulled out a map and unfolded it, until it covered his half of the windshield and instrument panel. And, without looking up, he began tracing his finger along our route of flight. He studied the map for several minutes. The fact that I had absolutely no flight training experience didn't seem to concern Tom in the least. At first, I just sat there wondering what I should do.

After a moment, I looked over at him with his head buried in the map, and then up at the airliner size instrument panel with what seemed like a hundred different gauges to be watched. With nobody holding the control wheel I was afraid that the airplane would fall out of the sky I quickly grabbed it with all my might. I thought that if I didn't hold the wheel tightly enough something bad would happen. I knew it had never occurred to Tom that I had no idea of what I was supposed to do.

'Just keep the wings level,' he said, casually; as he refolded the map and then turned it over and opened it to the other side. 'There's really nothing to it,' he added, without even looking up.

When it came to airplanes, Tom really believed that there was actually nothing difficult about flying. 'Just keep the 'blue side up,' he said, nodding towards the view through the windshield. For some reason he always trusted people.

•

Tom's greatest flying exploit almost cost him his life, and

occurred on a Saturday morning. Tom and I were working at Douglas Aircraft Company in Long Beach, and we were scheduled to work overtime one particular Saturday. The shift was to begin at eight in the morning. And at the appointed time, the entire technical staff was at their desks working, except for Tom. Tom wasn't there. And to say that his absence was conspicuous was an understatement, since he happened to be the department manager. So, that meant that everyone was there but the boss.

At ten thirty that morning, Tom finally showed up. When he walked into the Department all activity stopped. His appearance was dramatic. He was wearing a sheepish grin and dressed in a powder blue flight suit, with red piping down the legs and gold wings embroidered over his left breast pocket. The entire staff sat up applauding his entrance; some even whistled. He looked like a cross between a 'Hollywood' airline captain and a Bell Captain at a local hotel.

He wore an embarrassed look as he sauntered into the room as if he'd just landed his Boeing 747 at the wrong airport. And it wasn't until after he'd settled down and had a cup of coffee that he told us why he was late. And what'd happened to him had the office laughing for the rest of the morning.

•

It seems that Tom had spent the previous night with his girl friend at her apartment in Burbank. And, instead of driving the two hours in his car to her place, he decided to fly there in his airplane. It was very convenient for him to use his airplane, since Douglas Aircraft is located on the field at the Long Beach Airport, and his airplane was parked just a few hundred yards from his office. And the flight to Burbank from Long Beach is only twenty minutes, thus saving him a two hour drive.

After a late night with his girl, Tom woke up Saturday morning in her bed. It wasn't six thirty like he'd planned. It was eight thirty and he'd overslept and was late. He hadn't told his girl friend that he had to work Saturday; she was in the kitchen preparing a leisurely breakfast.

Frantically, Tom dressed and even helped his girl pull on a pair of jeans and a sweater, as he pushed her out of the apartment in his haste to get to her car. With Tom at the wheel of her convertible, they raced through the deserted streets of Burbank, heading for the airport. And as they drove along at high speed with the top down, the sun bathed everything in a bright light, as the day warmed. And as they neared the airport, Tom began to relax, realizing that it was going to be a beautiful day, and in a half hour he'd be at his office.

When he got to his airplane there was no time for a preflight 'walk-around' inspection. He just jumped in and started the engine. The engine of the TriPacer reluctantly coughed several times and came to life, as he pumped the throttle instead of taking the time to use the 'primer.'

Besides not having taken the time to preflight the airplane, he also skipped calling Flight Service for a weather briefing. Why bother, he thought, since the sun was shining brightly in Burbank. With Long Beach just thirty air miles away, what was the point? He didn't need a 'Weather Man' to tell him that the enroute weather was clear and that he'd have a wonderful flight.

Besides being a 'marrying fool' when it came to women, Tom's only other character flaw was that of procrastination. He never seemed to have the time to follow through with important things especially when it came to airborne safety. Tom's priorities often defied logic.

A year before, Tom had meant to refurbish the TriPacer's

instrument panel, installing new instruments and switches. But, after some while, he'd only gotten as far as pulling out the compass and his gyro flight instruments that were critical for instrument flight, as well as his electronic radio navigation instruments — the VORs. He did replace the whiskey compass with a new state-of-the art vertical compass, but after installing it, he never got around to calibrating it, or 'swinging it,' as the procedure is called.

Undaunted, Tom took off into the sunshine for the short flight to Long Beach. The TriPacer took off and climbed out over Burbank's Runway 15, heading south towards the hills of Griffith Park. Tom thought that even with a late start, the day ahead would work out all right. After all, what could go wrong?

It wasn't until the airplane had gained enough altitude to clear the hills of Griffith Park that Tom could see what could go wrong, and then some. As the airplane rose above the hills he realized that the entire Los Angeles basin, south of the hills was nothing but fog. During the night, a low ceiling of fog had drifted on shore, obscuring the surface from beyond Catalina Island to the southern slopes of Griffith Park. Of the southern California basin, only the San Fernando Valley was clear. And, as he flew onward, the Valley was quickly moving behind him, leaving only a carpet of fog below.

He sat at the controls, dumbstruck, his face frozen in disbelief. The TriPacer flew on relentlessly over the white blanket of fog, at one hundred and thirty miles per hour. The longer he flew the more serious his plight, as his immediate problem became the LAX TCA. His airplane was closing on the Terminal Control Area of Los Angeles International Airport, somewhere ahead.

Now, a Terminal Control Area, or TCA, is a positive control area in which an airplane cannot penetrate without prior permission from the authorizing agency; and for LAX's TCA, that agency would be the 'Los Angeles Approach Control.' The TCA is an area carved out of the airspace surrounding a high density commercial airport, to protect airliners from other traffic, as they approach to land or takeoff and climbout. And to penetrate that airspace without permission was an offense that could mean instant surrender of one's pilot's license — not a trivial matter to a flyer.

Tom had been flying his entire adult life and had never once lost his 'ticket.' And to compound things, he'd been a Certified Flight Instructor for more than twenty years, and if he now violated one of the Cardinal rules, the Feds would surely take offense–maybe take his CFI license forever. After all, a CFI teaches people how to fly and has to know all of the rules backwards and forwards. The bottom line was, he knew he just couldn't 'bust' the TCA. But staring below at the layer of fog, where exactly was LAX?

He sat there weighing his options. There weren't many. If he'd only finished the rehab on his panel or hadn't removed his radio navigation receivers, the VORs, at least he would then be able to locate his position. With this fog, the only way he could land at Long Beach was on instruments; and he didn't have any left in his panel. Instrument pilots are trained to fly and land with a partial panel of instruments, but he didn't even have enough instruments left for a 'partial panel.' How was he going to land at Long Beach?

He'd done the approach at least four hundred times over the years and it really was easy with the proper instruments; but, without them, it was impossible. And, in this weather, landing on instruments was the only viable option, other than turning around and going back to Burbank.

He was concerned but not worried, as over the years he'd been in life threatening situations before, where his airplane was close to coming apart, and still he survived. He thought that today's problem was not terrible, just sticky. Somehow, he had to figure a way through it without getting trapped by the Feds. The worst thing that could happen would be for the FAA to inspect his airplane. In its present condition they'd tear up his license. That is, if he was lucky enough to get on the ground without killing himself.

He had to do something, and soon, as the TriPacer was moving at over two miles each minute in the general direction of LAX, and according to the altimeter in front of him, the airplane was at three thousand five hundred feet. His eyes swept the remaining instruments in the panel; like his options, there weren't many. What remained were the altimeter, the airspeed, and the vertical speed indicators. Not enough to do much with, really.

He considered his options. Option one: continue towards Long Beach and try to fake some kind of an instrument approach with his limited instrumentation, or option two: return to Burbank, borrow his girl's car and drive the two hours back to Long Beach. He didn't like either option, and looked at his watch. It was already going on nine fifteen. If he drove to Long Beach it would take him until noon, and that would never do. He'd just have to be resourceful and go with option one.

Most, if not all Sunday pilots wouldn't hesitate about doing a 'one eighty' and return from whence they came, traveling from 'bad' weather into 'better' weather. But not Tom. He had so many flight hours under his belt and had survived so many bad situations flying airplanes, that unless the airplane was coming apart in flight there always seemed to be a way out. The fact that he couldn't see the ground and had no instruments for a 'blind' approach, was just something he had to resolve.

He actually told me one time that he knew in his heart that he wasn't going to die in an airplane, and therefore no matter what happened there would always be a way out. After he told me that, I guessed that about the only thing that could get his attention while flying an airplane would be a fire in the cockpit.

His right arm reached back, fishing for his flight bag on the rear seat. It wasn't there, and he half turned to look. But he always kept it there. The seat was empty.

Where could it be? He tried to remember. Then suddenly, he laughed out loud, remembering that he'd left it in his girl's car. He forgot to take it with him when he got to the airport. But, as he thought about it, not having it with him wasn't that big a deal, since he'd long ago memorized the approach plate for Long Beach's ILS Runway 30 Approach. He really didn't need it. All the 'numbers' were in his head. But, then he didn't have the instruments in front of him to see the 'numbers,' and that was a real problem.

His most immediate problem was the TCA; and he had to figure out geographically where he was–an easy matter if he still had his VORs, but the two vacant holes in the panel in front of him attested to the fact that, on this flight, Tom was never going to know his exact position.

The new vertical compass was now his only navigational tool, but it hadn't been calibrated so there was no trusting it, and no way of knowing how far off its readings could be.

So, he thought to himself, 'what the hell,' he'd try and see if he could land at Long Beach anyway. Fearless as ever, in typical Stansbury style, he picked up the mike and called Los Angeles Approach Control, deciding that his only chance was to fake it and shoot the approach without instruments and without the Approach Plates.

•

At best, in clear weather, an instrument approach is almost

impossible to execute without flight and navigational instruments and the appropriate published approach plate. But now, to attempt an instrument approach in real instrument conditions, on top of a complete blanket of fog thick enough to obscure the terrain inland right up to the mountain slopes, was not only insane but suicidal. Only John Wayne in the movies, or Tom Stansbury in real life, would attempt it.

'Los Angeles Approach, TriPacer 7622 Xray, request.'

'TriPacer 7622 Xray, Los Angeles Approach, say request.'

'TriPacer 22 Xray is a PA-22 slash U,' he said smoothly. 'We'd like an ILS into Long Beach this morning, and we're at three thousand five hundred, about ten south of ELMOO' intersection.'

'TriPacer 22 Xray, squawk 5623 and ident.'

Tom waited, knowing that he didn't have a radar transponder on board, and there was no way of 'squawking' any kind of a radar code. That was one of the 'black boxes' he'd removed and hadn't gotten around to reinstalling. He just had to play along.

After a pregnant pause the controller's words boomed in Tom's headphones. 'TriPacer 22 Xray, we're not painting anything at that location. Recycle your transponder.'

'22 Xray WILCO.' Tom responded, as calmly as he could. He realized that within a matter of seconds the jig might be up.

Another long pause.

'Negative contact, 22 Xray. Fly heading 180.'

'180 on the heading, 22 Xray,' Tom repeated, turning the airplane until the compass indicated 180 degrees. He hoped the compass wasn't too far off.

Thirty seconds went by and finally the voice returned, sounding irritated. '22 Xray, turn right to 270.'

'Right to 270 on the heading,' Tom replied. '22 Xray's transponder light is not blinking. I think it just died on me.'

'Roger.'

There was another long pause. Tom sat patiently waiting and listening to the sound of the engine droning onward.

'Radar contact! Fourteen miles west of Catalina Island,' the voice said, sounding even more irritated. 'Your heading is off by more than twenty degrees, 22 Xray. Turn left now to 080 before you end up in Hawaii.'

Tom would've snickered at that, but he realized that he was almost fifty miles out to sea, on top of the fog, and had no idea in which direction the compass was off. Approach Control was vectoring him into the final approach gate for Long Beach, which meant that when they'd turn him loose for the approach, he would be at the 'phantom point' one and a half miles from the Final Approach Fix, in line with the runway. But mental visualization of an 'electronic fix' and situational awareness is everything in instrument flying, and Tom hadn't a clue as to where he was or where he'd been, and had no visual reference with the ground to know which way Long Beach Airport was.

All he could see in all directions was a white carpet of fog and, the ranges and peaks of the inland mountains.

'22 Xray, are you having equipment difficulties?' the voice finally asked, sounding frustrated.

These were the words that Tom had been waiting for, and suddenly his spirits lifted as he keyed his mike. 'TriPacer 22 Xray is declaring an emergency. I've just experienced complete electrical failure and would like a 'No Gyro Approach' to Runway 30.'

'Roger 22 Xray,' the voice said in a relieved tone, now that there was a plausible explanation for Tom's erratic maneuvering.

And so, Tom was talked down to a landing at Long Beach. And once on the ground, after a lengthy conversation by phone with the watch supervisor of Los Angeles Approach Control, he once again was able to charm his way out of another 'tight situation.'

•

The moral is, sometimes more flight hours in a log book are just more flight hours. They don't mean that the pilot is necessarily safer.

As long as the FAA uses flight hours to indicate a pilot's experience, then, bigger numbers signify better judgement and safer flying. This may be true with commercial flight crews, but I'm not betting the rent money on private pilots using the same criteria. I've just known too many high time pilots that seemed to have a 'death wish' when it came to their flying judgement.

In Bob Johnson's case, he proves that low numbers in one's log book indicate lack of experience and prove FAA's point.

In the end, I guess the trick is to gain the hours and the experience, without losing respect for the danger, becoming complacent to that danger, just because poor judgement hasn't caught up with you yet.

Chapter Two
The Always Dangerous Student Pilot

Solo Adventures of a Fledging Flyer

The two words that can turn a 'high time' veteran pilot's blood to ice, and fill him with terror are the words 'Student Pilot.' Especially if he hears these words over his aircraft radio while approaching the same airport. And, if given half a chance, the veteran, realizing that 'discretion is the better part of valor,' will immediately depart the area and circle at a high altitude some distance away until the student is safely on the ground.

Student pilots are very much like student drivers behind the wheel for the first time on neighborhood streets. They're a threat not only to other drivers, but to anyone walking on the sidewalks, as well. And, like their landlocked counterparts, student pilots offer the same danger to everyone in the air; their behavior and reactions to normal aeronautical situations can be totally unpredictable.

The following story describes my tenure as a student pilot, and in part explains why student pilots everywhere are given such a 'wide berth.'

•

I guess all Sunday Pilots daydream about flying when they're not in the cockpit of an airplane. I know on long freeway drives I tend to lapse into the left seat of a single engine Dehavilland Beaver flying the desolate stretches of the North. At other times, I'm flying a DC-2 airliner at three thousand feet, in a blinding snow storm, somewhere between Buffalo and Cleveland, while trying to keep the airliner aloft, with three inches of ice already on the wings.

So, it seems natural for any pilot to occupy his thoughts about flying when he's 'grounded.' But, for recreational pilots who don't fly for a living, fantasy flying during the week, when we're busy working for a living, is pure escape.

I've been crazy about airplanes ever since I saw my first one. I was four years old. It was a yellow Piper J-3 Cub and it flew past me over a clearing of a woods. It was very low, just a few hundred feet in the air. It was love at first sight. And it was because of this love affair that I did so poorly in Mister Crookshank's Algebra I class in junior high.

I was fourteen and my 'hormones' were boiling over. When I wasn't fantasizing about girls, I was thinking about airplanes. My junior high had two lunch periods back to back. And it turned out that my Algebra I class was held during the first lunch period. So while others had eaten and were outside having fun with their friends, I was sitting in the back row of Mister Crookshank's second floor classroom, next to a half open window.

From the back row, laughter and yelling wafted in through the window from the yard below, making it impossible for me to concentrate on solving linear equations. And to add to my lack of concentration, my stomach was growling from hunger.

To keep my mind off my stomach and algebra and girls, the

only thing left to think about was airplanes. And, I'd usually fantasize that I was piloting a B-17 Bomber during World War II — sometimes over Germany, sometimes over the oil fields of Romania. I can't tell you how many missions I flew that one semester, but it's got to be pretty close to the actual number of sorties flown by the U. S. Army Air Corps during the war.

All through my childhood and into adulthood, airplanes have been a large part of my 'thought life.' And from childhood, one of my lifelong goals was to someday become a flyer. For some reason, I thought that flying an airplane through the cloudless skies would be the most thrilling thing a person could ever do. And the thought of it sustained my interest in flying until a time when I was actually able to take flying lessons and get my pilot's license.

So for me achieving my Private Pilot's License became a 'right of passage,' an important milestone. And, just the accomplishment of the thing made me feel special. The investment of so much time and money and effort was a substantial sacrifice for me; but once realized it seemed to make me a 'whole' person. The license alone was of great value, because it symbolized both my sacrifice and achievement. And, at that moment, I knew I could die a happy man. I knew that no other achievement in life would mean as much. And, with that in my wallet, I felt special.

•

Like most everything else in life, it was through a meandering series of events that evolved over a number of years, that caused me to find myself at the Long Beach airport, at a Fixed Base Operator called 'C&J Air.' After a lifetime of waiting, I was finally ready to begin my flight training.

A Fixed Base Operator, I discovered, was what flight schools were called. FBO's are the one's that rent airplanes to licensed pilots, as well as to student pilots with their flight instructors, and they make their money through the renting of the airplanes, like Hertz does with cars.

C&J Air was a rundown 'hole in the wall' operation, in a remote corner of the Long Beach airport. It's office building was a Quonset hut from World War II, when the airport was used as one of the many Army Air Corps training centers.

I met a handsome young man by the name of Bill Nielon, who was in his early thirties. He was introduced as 'C&J's best flight instructor.' And if image equates to competence, then he was the best. He looked like Madison Avenue's idea of an airline pilot. If he'd been in uniform he would've made a great magazine ad for American Airlines.

He wasn't too tall, too thin, too athletic or too tan. Plus, he had a certain boyish wholesome look about him. And when he smiled all you could see were his beautiful white and straight teeth. His smile with his sandy blond hair gave him a 'Tom Sawyer' quality that immediately put me at ease. He exuded trust. If you couldn't trust him, you couldn't trust anybody else in the whole world.

And when he talked, his soft-spoken manner and his low key approach gave me the impression of total sincerity and confidence. His manner reminded me a lot of Will Rogers.

Bill seemed to be the sort of teacher that would always keep his cool, and when I thought about other attributes of the perfect flight instructor, Bill's demeanor immediately came to mind. And, over the next five months, and seventy-five flying hours, my initial impression of Bill proved to be correct. He was as advertised; what you saw was what you got. And to this day, I've never met anyone as cool and calculating and deliberate as Bill Nielon.

For the next five months Bill taught me all of the rules, regulations, and flying techniques required by the FAA curriculum for the Private Pilot's license. It was not a romantic time for me, as the fantasy of training was quite different when measured against the reality of my talent and capabilities.

When you fantasize about how a thing will occur, like flight lessons, and then experience the actual lesson– there's a world of difference. My skill and performance always fell short of my fantasy and personal expectations. That's the problem with fantasies–they never measure up to reality.

•

Anyway, as a part of the latter stages of flight training, the student pilot is required to perform a series of cross country flights, first with his instructor, and then on his own. I suppose the reason for that is to confirm that if a student can fly solo from point A to point B and return without killing himself or hurting the airplane, then he's competent enough to complete the remaining curriculum. Of course, with death as the alternative to sloppy flying skills, the student is motivated to concentrate and perform well. I've always looked at the solo cross country flights as the FAA's way of 'thinning the herd' of incompetent pilots. And, if the pilot survives his solo cross country trips, then he can probably pass the FAA checkride for the license.

Although there are several cross country flights required of the student each of varying distances, the 'long cross country' is suppose to total 250 nautical miles; meaning that the outbound and return legs should be no more than about 125 miles each. A flight of any greater distance is really too much for a student to handle with his limited experience. The total time for this solo trip should be in the range of about three to four hours.

Even though my flight experience was only fifty hours at the time, ten hours of that were dual cross country. Those ten hours gave me a false feeling of confidence, and for some reason I got it in my head that I wanted to fly from Long Beach to Phoenix and back for my long solo cross country. A trip of that magnitude totaled 640 miles, two and a half times longer than that required by the FAA.

The following flight to Sky Harbor Airport from Long Beach and back, is proof that the FAA knows what it's talking about when assessing a student pilot's ability to cope with the inherent dangers of an extended flight.

The Flight To Sky Harbor

It was late April, 1973 and my flight training was almost finished. It was now time for my long solo flight. And, hopefully, after that was accomplished, all that remained would be to polish up my skills for the FAA checkride.

One day, without warning, Bill Nielon was suddenly gone. He never even gave me a hint that he would be taking a month off to attend a four week Flight Engineer's school at the Flying Tigers Airlines facility at LAX. I knew he was trying to get on with the airlines, and I guess he figured with the Flight Engineer's written exam already passed and behind him, he'd have an edge when interviewing with the airlines. At the time, I thought that the least he could've done was to tell me of his plans.

It was during Bill's time away at Flying Tigers that he and I were down to just one flying lesson per week. That was so I could fly extra solo time during the week prior to the checkride. And now, since he wasn't going to be around for a month, I just assumed that he expected me to complete my long solo cross-country. I assumed he expected me to be ready for a final review phase of training when he returned.

The distance from Long Beach to Sky Harbor Airport in Phoenix and back is 640 nautical miles, too far really for a student to fly solo, especially in a Piper Cherokee 140, whose cruise speed is only about 100 knots at best, even in a 'no-wind' condition. That meant that it would take about three hours and twenty minutes plus another ten minutes for 'climb to cruise altitude,' or about three and a half hours each way for the flight, if everything went right, and no problems developed.

To fly there and back in the same day is too much for a student pilot to handle. The total flight time is too taxing,and when a pilot is tired and can't 'stay ahead' of the airplane, that's when things happen–things that, to an alert pilot are routine, but to a tired pilot with slow reactions, could cost him his life.

I know that if Bill was there, he would never have let me fly solo to Phoenix and back. And, I'd like to think that most Flight Instructors wouldn't either, but C&J Air was comprised of a motley crew of instructors that never seemed to worry about the safety of their students. Bill Nielon was the only one who rigorously adhered to and taught the principles of safety. I didn't realize it at the time, but I would have to get a couple hundred flying hours under my belt before I would come to believe that a pilot should pay attention to his personal limitations, as much as to any other margins of safety. But undaunted, I pressed on with my obsession to fly to Arizona.

The reason I wanted to fly to Phoenix in the first place was that my wife Barbara and I had friends there. And, my fantasy throughout my flight training was a scene in which I tell her that we could fly there by ourselves, rather than drive. Since she had no idea that I was taking flying lessons she'd be totally surpprised. My rationale was that if I had already flown the trip while a student, then it should be an easy matter to repeat it after I was licensed.

Looking back on it now, all I can say in my own defense, is, it just seemed like a good idea at the time.

•

No matter the age, the blessing of being a student is in having ultimate confidence, even though there is no experience to back it up. Students have a fearlessness that is unrivaled; they don't clutter their minds with a lot of facts concerning the realities and dangers of aerodynamics and physics. All they know is that when they push the control wheel forward, 'the houses get bigger,' and when they pull it back 'the houses get smaller.'

One of the things Student Pilots have in great abundance, short of a Guardian Angel, is luck. And most of them need much more than their fair share to survive.

I must have been destined to fly my Sky Harbor trip, since, as luck would have it, on the day of the flight, the only CFI available at C&J to approve of my trips was someone I'd never even seen before. It turned out that he usually worked only on weekends, but he was substituting that day for an associate who was sick. I knew that unless my log book was endorsed by a CFI, my flight would never happen. I was very apprehensive about it, as I'd already taken a day off from work.

As I look back, there was never any reason to be apprehensive, knowing the mentality of the C & J instructors. The staff CFI's were rather cavalier about safety issues; as CFI 'Captain' Richard Graff would say, 'Heck, If ya run into any trouble, just land somewhere.'

But requesting a stranger's approval for an undertaking that even I knew was foolhardy worried me. But true to the spirit of C&J, the CFI came through for me, as he paged through my log book, and then finally stamped his endorsement for the flight on its inside back cover.

'Have a good flight!' was all he said to me, handing me back the book, and that was that. And, a moment later, I walked out into the sunshine of the flight ramp, book in hand, and began what was to be a very memorable adventure.

•

The day was clear with a few high clouds that lay scattered along the coast from Santa Barbara southward. I was told by the weather briefer on the phone, an hour before, that weather would be clear all the way from California through Arizona. I'd filed a VFR flight plan from Long Beach to Sky Harbor and all was in readiness. I had 'planned my flight' and now I was going to 'fly my plan.' Life was good! What could possibly go wrong?

•

It didn't take long for fate to intervene. Actually, I'd just taken off from Long Beach, and was climbing to the cruise altitude of nine thousand five hundred feet, a Herculean feat for a Cherokee 140, with its one hundred and fifty horse Lycoming engine and fixed landing gear. For those who have never had the experience of flying a totally under-powered airplane like the Cherokee, the climb to altitude usually takes three days, and by the time you're at cruise altitude, it feels like the trip ought to be almost over.

By the time I finally reached altitude, I had everything in the cockpit set and under control. I even had the airplane trimmed up for 'hands off' flight, although that was not a possible 'real life mode' in a Cherokee. After finishing the pilot chores, I radioed Flight Service to activate my flight plan. And, it was then that I learned how quickly a weather forecast can change.

As I talked to Los Angeles Radio, the call sign of the Los Angeles' Flight Service Station, the airplane was humming along nicely, tracking the 238 degree radial of the Paradise VOR. As I activated my flight plan to Sky Harbor, I asked about the weather through the Banning Pass, which I'd reach in about a half hour.

During an earlier weather briefing there was no mention of a dense deck of clouds that suddenly appeared in front of the windshield. The weather was supposed to be clear from Long Beach all the way to Phoenix. So how could the weather change so suddenly? Student Pilots, didn't expect things like that to happen.

As I listened to the overhead speaker, the 'revised' forecast was passed on to me. And as the voice talked, the airplane suddenly became engulfed in total whiteness; clear one second and totally opaque the next. It happened in an instant and I was almost overtaken with panic.

For several seconds I just sat there paralyzed, not realizing what was happening and not able to explain why I couldn't see out the windows. It was like flying through the inside of a bottle of milk. Then after a moment it dawned on me that I was still very much alive. And I also realized that the airplane was still flying. Then it hit me that I'd flown into a cloud. I couldn't see anything outside, but I also knew that I had to control my fear of not seeing outside the airplane, and that I had to relax, turn the airplane around and fly out of the cloud.

That was easier said than done. My heart was still racing, and I was sure I was about to crash and die. But, as I began to relax, I did as Bill had taught me. First, I began a serious instrument scan with what limited training I had. Then, I banked the airplane into a constant rate turn to the right as my eyes kept moving between the miniature airplane of the attitude indicator and the sweep second hand of the clock.

After exactly sixty seconds of maintaining the airplane's attitude on the 'artificial horizon, 'the airplane had turned one hundred and eighty degrees. And, once that was accomplished, I rolled out, and after several moments the airplane flew out of the 'whiteness' and into the clear.

As the airplane turned slowly through the whiteness, I was amazed to think that I never saw the cloud before it engulfed the airplane. The air around me seemed clear for some distance. There were some clouds in front of me, but they seemed to be miles off. Then suddenly everything was white!

After I recovered from the maneuver, I wondered whether the cloud that I'd encountered was an Omen. That maybe I was tempting fate with such a foolhardy flight. Flying into that cloud was really intimidating. I'd never been inside one before, and it was doubly frightening, since I was alone at the controls, and had no idea of what to expect. I wasn't prepared to deal with it, and thought I was pretty lucky that I didn't panic to the point of losing control of the airplane.

As I droned along in clear air, I thought about all of the pilots who get their instrument rating without ever flying through the inside of a real cloud. And here I was, stumbling into one without even knowing it. I felt lucky that I survived.

The revised forecast recognized that the weather was changing, but was only supposed to be a problem from the Ontario area eastward through the Banning Pass. The voice on the radio told me that if I could make it somehow to Palm Springs, and get east of the mountains, then the weather would be clear all the way to New Mexico.

As I backtracked towards Long Beach I thought about how I was going to get to Palm Springs without going through the Banning Pass. I sure didn't want to fly into anymore clouds. So after a moment, I decided to go south along the coast down to Oceanside, where I could turn inland and cross the mountains at the Julian VOR Station, and pick up my route again at Thermal, thus cutting off the segment from Banning to Thermal.

'What are your intentions, 69 November?' the voice asked over the cockpit speaker.

'This is 69 November, and I'd like to amend my VFR flight plan routing from Long Beach to Sky Harbor. Change the routing to read 'via' Oceanside, Julian VOR, Thermal VOR, Victor 16, direct Sky Harbor. Over.'

'Roger, 69 November. We'll make the change. Anything else I can do for ya today?'

'Negative,' I replied. 'Just extend my ETA by forty five minutes, and thanks.'

And with that I turned the Cherokee southward following the coastline. But, as the airplane and I flew south I could see that a low bank of Stratus a half mile or so off shore would soon be drifting inland, obscuring the coast. It was suddenly obvious that later in the day, when I'd be returning from Phoenix, the coastal route would be nothing but fog. If I wanted to return to Long Beach it'd have to be via the Banning Pass. I didn't feel good about that prospect, but I was too busy at the moment to care.

I should have been concerned about the return trip right then, but things were happening too fast: the clouds and the change in the weather and now the fog. But, like all students, if a decision about something didn't need to be made right then, then I didn't have to think about it until later. The fog would be a problem for the return trip, hours from now.

Some twenty five years later, as a 'Monday morning quarterback,' what I should have done right then was abort the flight, especially after seeing the fog drifting in along the coast. With all the unplanned things that were happening to me I should have waited another couple of weeks, when the weather would be better for the trip. But, like all students impatient to implement their training, I was anxious to get on with the things that had to be completed prior to the checkride. As I flew along, something in the back of my mind told me that if the weather is worsening on the outbound leg, it surely isn't going to improve six hours later for the return flight.

And, like all Student Pilots, I didn't worry about return legs while still trying to fly the outbound one.

•

Forty minutes later, I was flying over the Julian VOR station on top of the mountain. I was at eight thousand five hundred feet, as I cleared the peak staring down at the tiny 'lighthouse' shaped VOR station. And as the airplane flew beyond the summit, the day seemed to brighten and I could see blue skies and bright sun ahead for what seemed like hundreds of miles. With better weather in front of me, what could possibly go wrong?

To navigate at this altitude, all one had to do was pilotage. My Sectional was in my lap, and my right index finger crept along the plotted course line as the airplane slowly advanced over the earth. From this height the VOR reception for radio navigation seemed to be fine. And as I looked down to my left, I could see the green area south of Palm Springs that would be Thermal. And looking off to my right, I could see the north shore of the Salton Sea. The view was spectacular, a view only pilot's of light airplanes would ever see.

The view was so intoxicating that I allowed myself a precious moment just to revel in it and the pure joy of flight. And at this moment, flying above the earth, a 'regal' feeling came over me–and with it, a certain peace. It was a good feeling. I felt separated from my cares by a couple of thousand feet and a million miles, Amelia Erhart wrote about a similar feeling in her book, "The Fun of It,' where she describes her feeling of the joy of flight.

I could only imagine how much better the feeling would be after I'd have a license and could jump into an airplane any time and go flying. This special feeling that flying gave me is why I sacrificed all my time and money for flight training–so I could have this feeling forever.

•

A half hour went by, and then another. I was getting close to the Colorado River and the town of Blythe. That's the point where I had to switch fuel tanks. It was pretty close to half way. As the minutes passed, the day seemed to improve greatly, once I was east of Julian.

The sun was shining brightly and the sky was totally clear, except for some high Cirrus. And, as I passed over the Blythe Airport, I switched fuel tanks. It was an easy procedure: Auxiliary Fuel Pump to the 'ON' position; then move the handle of the Fuel Selector Valve from 'LEFT' to 'RIGHT.' Wait a few seconds, then turn the Auxiliary Fuel Pump to 'OFF.' Done.

It was amazing just how simple flying an airplane really was, as long as everything worked and nothing out of the ordinary happened. It's mostly the flying environment that complicates things, especially in a metropolitan area. I think Bill Nielon told me once that fifteen percent of the total number of airplanes flying in the United States at any one time are airborne in the skies over Los Angeles.

Thirty minutes east of the river, the brightly shining sun dissolved behind a cloud. I couldn't believe what happened. In an instant, the light intensity was dimmed from bright and clear, to dim not so clear. As a result my confidence waned with the light. I wasn't worried or anxious, but the light had buoyed my spirits and now my spirits and my faith in my flying skills were at a low ebb.

Looking up through the top of the windshield I could see that the sky was clouding up. And as the ambient light diminished even more as a high overcast developed, my general attitude and outlook changed as well: from a good feeling to a feeling of uncertainty.

I was still thinking about the cloud layers when a sudden movement in the instrument panel caught my eye. It was the bobbing of a warning flag on the number one VOR indicator. I checked the chart in my lap and discovered that I must be in the marginal reception area, halfway between two VOR stations; between Blythe and Buckeye. I checked the altitude. At my present cruise altitude the reception range of 'low altitude' VOR stations was only forty miles. And at that moment, Buckeye was too far ahead of me for its signal to be usable, and I was also too far beyond the river from Blythe VOR for its help.

Just as I looked up from the chart, the warning flag on VOR number two was also telling me that the signal from its station was now too far away to be of use. Now, I wasn't sure what to do. This situation had never come up in my training. At first I thought I could triangulate using other stations off to each side, but one glance at the map told me that I was now in an area where any two stations were about a hundred miles apart. This was one situation that Bill hadn't prepared me for.

Without a clue as to what to do, I decided to just hold my course and look down at the terrain for some features I could identify from my sectional chart and use for pilotage.

Interstate 10 was a highway that happened to be about ten

miles off to my left and was angling northeasterly from my heading. And, since it traveled all the way to Phoenix, after a fashion, I decided to follow it. After all, with the road below, using pilotage, what could possibly go wrong?

•

Interstate 10 meanders easterly from Los Angeles through San Bernardino, the Banning Pass, and into Palm Springs. From there, it turns southward to Thermal, and then easterly again to Blythe and the Colorado River. From the river, without any curves, it goes straight to Phoenix. There is one problem with its route, however. It has a slight jog to the Southeast at a point fifty miles east of Blythe, or about forty miles east of the river. And as it happens, the geographical bend in the highway is directly below the marginal reception zone for the Blythe and Buckeye VOR stations; exactly halfway between them.

It would be difficult indeed to have planned this coincidental phenomenon. But it happened. And although I learned later how to accommodate this navigational problem, it wasn't until long after suffering the consequences of my actions.

I had folded the Sectional to a convenient length for my lap board. And now encountering the 'dead zone' of VOR reception, I was also about to fly off my map. So to unfold the map, I let go of the control wheel and opened the map to the other side, something that Tom Stansbury could do with aplomb. But, as I did so, the remainder of the course line was unveiled from the River to Phoenix. I slid the map under the spring clip of the lap board and studied it.

As I let go of the control wheel, the airplane's left wing began to dip ever so slightly, because it was heavier with me sitting in the left front seat. And, as the seconds passed and I occupied myself refolding the chart, the angle of bank continually increased until the engine noise took on a subdued high pitch scream. The sound took me by surprise.

Now, whatever procedural training my curriculum lacked, the recovery techniques for unusual attitudes wasn't one of them. And I knew in a heartbeat the unmistakable sound of a 'Graveyard Spiral.' But, by the time I got myself untangled from the chart and looked up through the windshield, the airplane's nose was approaching a dive–straight down.

Quickly, I pulled the throttle back to idle and leveled the wings, then after several seconds, very gently eased the control wheel back with both hands, careful not to over stress the wings on the pull out, as the airplane's nose slowly began rotating upwards toward the horizon.

By the time it was over, my heart was pounding. In real-time, it probably lasted less than twenty seconds from start to finish, but from where I sat, it was a nightmare-like happening in slow motion. Like being chased in a dream by a monster when your legs can't move.

As if that spontaneous aerobatic maneuver hadn't been emotionally stressing enough, when I looked down from the side window at the highway, my heart stopped. I was shocked. The highway was gone. Disappeared, as in 'not there anymore.'

Again, strange things were happening. And they were happening so fast I could not keep up. I looked out the windshield for the highway–nothing. Only the vastness of the open desert of Arizona lay ahead, the endless sight of undisturbed wilderness, with no signs of life. Now I was in big trouble and knew it. My heart started pounding again, as fear of being lost welled inside me. In an instant panic was flooding through my mind in waves.

I looked out of the right side window–no road. What was going on? How could a major Interstate highway disappear in less than a minute? In my emotional state nothing was clear to me. I was in no mood to try to figure things out. That would have to wait until later, when I was on the ground.

It's funny, but a two lane divided cement Interstate as wide as it is on the ground, with its center median of dirt separating one side from the other, should be an easy thing to spot from the air. But the truth is, even though it measures some thirty yards wide on the ground, from a mile and a half overhead, it's just a thin pencil line drawn through the open desert.

I got lost In less than a minute, after navigating for more than three hours and some three hundred miles from Long Beach. I now had no idea of where I was. And an ache began to grow in the pit of my stomach. The same feeling a child gets, when he thinks he's lost after being separated from his mother in a department store.

'Dear God,' I thought. Here I am just forty five minutes from Phoenix and I'm living a 'terror nightmare.' As the fear built up, my only hope to keep it from turning into all out panic was to come up with a plan, any plan. Something that I could execute and busy myself with that might get me out of this 'crack.'

Rationally, it was almost impossible for me to believe that this craziness was happening. I knew I was getting tired, and that a trip of this length was probably not smart. But I never thought that so many 'life threatening' things could go wrong.

Turning the page of a map was such a simple thing. I still didn't understand what had happened. But, I had more important considerations at the moment. The one thing I did understand very clearly, as a flyer, was the Lycoming engine sitting in front of me. It was consuming fuel at the rate of one pint every minute. I had to do something to remedy my situation, and fast.

•

The Arizona mountains are unique as geographic topography goes and makes navigating by pilotage easy. Arizona had mountains sprinkled everywhere, with distinctive peaks sticking up in all quadrants above the flatness of the desert floor. How hard could it be to compare the geography that I saw below to that depicted on the chart in my lap." It should be a 'no brainer."

My heart began to slow down a little as I worked out my plan. I was going to head eastward and pick out recognizable features of the landscape and compare them to my map. That way I could pinpoint my position. It was a good plan. Nothing could possibly go wrong?

I looked down at the landscape below and saw a narrow valley between two parallel mountain ranges. Then I looked at the map. There it was. According to the map, the peaks of the mountains were over five thousand feet high, emerging above the desert floor like a giant outcropping that laid in a northeasterly direction for the next forty miles.

This seemed so easy. And after a few minutes of comparing terrain features and assuring myself that I was following the correct range of mountains, my whole body relaxed. For the next forty minutes I followed the range below. And, according to the map, it would lead directly to Phoenix. And, as my finger traced the airplane's progress an inch at a time across the map, the minutes passed and I enjoyed the ride.

As the airplane approached the terminal end of the mountains, where the range flattened into what should be a large basin with Phoenix at its center, I put down the Sectional and strained to look into the distance for the half dozen skyscrapers that comprised downtown Phoenix.

I sat up in the seat and strained to look over the Cherokee's nose at the hazy horizon. For several moments I searched the distance for a faint glimpse of Phoenix. Finally, I realized that there was nothing there but more open desert. There was no Phoenix.

I couldn't believe it. I was totally caught off guard and shocked. I tried to be so careful in my pilotage, tracking the terrain features of the mountain range and verifying them on my Sectional. And it took a long time as the airplane flew along over what was supposed to be the outskirts of Phoenix before I realized that I'd been wrong. That I'd followed the wrong mountains. My high spirits were instantly dashed. And this time fear and panic that I'd suppressed for so long, suddenly surfaced; flooding my consciousness.

I knew that I was going to die. It was just a matter of time, after the fuel was gone. It would be in the open desert where no one ever goes. It would take years to find my body. My mind kept thinking about my eminent demise. I knew what the newspapers would read: 'As a Student Pilot of little flying experience, he should have never attempted such a long solo flight, especially over a wilderness area of several hundred miles, in a single engine airplane.'

I couldn't stop thinking about it. I'd followed the wrong mountain range! Not only that, but I'd wasted forty minutes of precious fuel. Fuel that kept me aloft and away from the hostile desert below. The forty minutes of fuel I'd just wasted was my reserve fuel in case of emergency.

Now my situation was totally clear to me. If I couldn't find Sky Harbor from my present position, wherever that was, then without the extra forty minutes of reserve fuel, there was no doubt where I'd be landing, and coyotes and rattlesnakes would be my only company, if I survived.

It was crisis time and I knew it. If I was going to survive this flight as a Student Pilot, I had to be rational and do things by the book. I knew now that I couldn't do it on my own; I'd already proved that. It was time for me to rely on others to save me.

Psychologically, I'd gone from calm through all of the gradations of fear; from annoyance to almost life-threatening 'fight or flight' feelings of panic. Emotionally, I was a wreck. I'd tried unsuccessfully on my own to locate Phoenix. I'd failed miserably, and only succeeded in using up my reserve fuel.

Mentally, I was exhausted and beyond thinking. Maybe years from now, after hundreds of flying hours later, I'd do better, but today, as a student, I really needed help. In my present state, the best I could reckon was that I had maybe an hour's fuel left. Enough to make Sky Harbor, maybe. But 'maybe's' weren't what you wanted to bet your life on.

It was time to do what Bill Nielon had drummed into my memory. 'If I was ever in trouble in an airplane, all I had to remember were the 'four Cs.' 'Climb, Communicate, Confess, and Comply.' And now, I was in trouble and it was time to see if the four Cs really worked.

I glanced down at my lap board and found the radio frequency for Phoenix Approach. I set it into the radio and picked up the mike.

'Phoenix Approach, Cherokee 8369 November. Over!'

There was no reply. I double checked the frequency on my chart. It was correct. There was a long moment of silence before I heard some traffic. It was an airliner.

'...Left to 230, down to three thousand, cleared for the approach, American 7720.'

Relief. At least I knew I was on the right frequency. I took a deep breath and keyed the mike again.

'Approach, Cherokee 8369 November, Over.'

'Cherokee calling, say request,' the voice demanded impatiently.

'This is Cherokee 8369 November. Student Pilot. I'm a PA-28-140, on a solo cross country from Long Beach to Sky Harbor, and I'm lost. Over!'

'Cherokee 69 November,' the voice returned in a much friendlier tone. 'How much fuel do you have on board and what's your altitude?

'69 November has about an hour's fuel, and I'm at eight thousand five hundred feet. Over.'

'Cherokee 69 November, are you transponder equipped?'

'Affirmative.'

'69 November, squawk 5127 and Ident.'

I wrote down the transponder code on my navigation log, so I'd remember it, and then dialed in the numbers and pressed the 'Ident' button.

A minute went by, and then another. Finally the voice returned.

'Cherokee 69 November, radar contact, forty eight miles northwest of Sky Harbor. Turn right immediately to a heading of 130. You're presently in the middle of an MOA that's hot! Descend and maintain four thousand, report leaving six thousand.'

'Roger Phoenix,' I said, writing everything down on my lap board as he spoke, and then read it back to him. 'Right turn to 130 and down to four thousand, and I'll report passing through six thousand, 69 November.'

I banked the airplane to the right and pulled the throttle back a little as I pointed the nose down. The rate of descent was one thousand feet per minute on the vertical speed indicator. As the airplane began descending, I looked at the Sectional to find my position.

I approximated fifty miles distance with my thumb and index finger and measured from Phoenix to the northwest. Sure enough, I was right in the middle of the Luke Air Force Base transition training area, where beginning fighter pilots learn to fly jets.

I just sat there shaking my head in disbelief. My mind reviewed the events of the morning and I wondered how I could've ended up here, so far off my course? Everything was fine until that graveyard spiral.

Phoenix Approach Control vectored me several more times, until finally I was on a five mile final to Runway 8 Right. But as I was handed off to the Tower, I thought I was much too high and was about to mention that fact to the Controller, when he called me with a request that solved my altitude problem.

'Cherokee 69 November, there's a Boeing 727 three miles behind you for landing. Can you give me a hundred and forty knots to the field?'

'I'll try,' I returned, thinking that to do that I'd have to practically dive the airplane straight down. And that's pretty much what I ended up doing, leaving the throttle at cruise and pointing the nose at the runway, while watching the airspeed needle move way into the yellow arc. And, needless to say, that was my glorious arrival at Sky Harbor — like Torra, Torra, Torra!

•

Once down, with Ground Control's help, I taxied to the Executive Terminal on the south side of the field. And as I was shutting down in a transient parking spot, an electric golf cart drove out from the Terminal to meet me.

'Do ya need any fuel? a middle aged man asked, stopping the cart in front of the airplane.

'Sure. Top both tanks and add 50 weight oil, if it needs it,' I answered, locking the airplane and carefully walking to the cart and gently sitting down next to him.

'Need anything else?'

'Ya, a rest room, and fast! I've been flying for more than four hours and I'm afraid to make any fast moves.'

'I know the feelin.' Fix ya right up,' he laughed, as we drove back to the Terminal.

After a half hour's rest, a candy bar and a soft drink, I was ready to sit down and try to figure out how I'd gotten lost.

Outside the terminal, the skies were getting more overcast. The wind that had been light and variable when I landed was now coming in gusts. About the only weather that hadn't changed was the air temperature — still a comfortable seventy five degrees.

I sat down on a bench outside the entrance to the Terminal and took out my Sectional. I opened it up all the way. And as I smoothed out all of the folds I instantly realized what had happened to me and laughed out loud. If I'd opened the map all the way during the flight, I never would've gotten lost. The way the Sectional was folded, it was easy to lose the interstate highway.

The highway turned abruptly twenty degrees, veering southeast, exactly at the fold in the map, and exactly at the same geographic spot where I'd performed the disorienting 'graveyard' spiral. And, it was during the spiral when the airplane's heading had changed. It went from 076 to 020 degrees, as it again recovered to straight and level. From a student's point of view, that's like changing directions from East to North.

So, when I looked out the window to the left, for the highway that had been running in an northeasterly direction, it was gone. I should have looked to the southeast instead, where the highway had changed directions, abruptly. The fold in the chart, however, didn't allow me to see that the road had turned, either before or after I refolded it. And the other mistake I'd made was not accounting for the new heading from the graveyard spiral recovery. Originally, my heading was East, but after the spiral the airplane recovered to straight and level in a northerly direction. So, when I thought I was still looking to the East, I was actually looking towards North.

As I sat there staring at the Sectional, I wondered if I should do anything in the way of flight planning for the return trip. The California weather was probably worsening by the minute, so I decided not to waste any more time, and skipped a weather briefing from Flight Service. After all, I knew what the weather was and chances are, it wasn't going to improve. I was looking at another four hours of flying, and there'd be plenty of time for me to think about the Banning Pass and formulate another plan of action.

Maybe as a student pilot, you just had to fly a flight and not worry about the consequences too much. If you worry too much, you'd probably scare yourself into fear of flying.

•

The Return From Sky Harbor

It was now 2:30 p.m. With overcast skies and a dim afternoon light, I knew that even if I started back this very second, the last half of the flight from Palm Springs to Long Beach would be logged as 'night,' There was just no way I could make it back before dark.

Actually, with all of my preflight planning, I figured that a flight in a Cherokee 140 would take only three and a half hours to Phoenix, when it really took closer to four. And now, going west, if there was any head wind at all, it would add another half an hour.

By the time I cleared the Banning Pass, back into the outcroppings of the Los Angeles basin, I was afraid that fog would be the problem, not whether or not there'd be any, just how much.

I got up and walked into the FBO and paid my fuel bill and took the receipt so I could get the fuel cost credited against the airplane's rental fee. I was in a hurry and didn't wait for the golf cart to drive me to the airplane; I needed to get out of town, and fast.

In the interest of time, I didn't bother filing a flight plan, and I decided to radio Flight Service during the return flight for the current Los Angeles weather.

The flight back to the river was uneventful, but a half hour west of Blythe, just as I was approaching Chiriaco Summit, the sun faded, setting beyond the mountains ahead. And by the time I flew another fifteen minutes, the twilight faded to darkness, as I arrived over Thermal and headed northwest towards Palm Springs and the Banning Pass.

As the twilight faded, I turned on the red cockpit floodlights so I could see the instruments. The airplane didn't have post lights on the individual 'big six,' so in the darkness, the view of the flight instruments was marginal at best. Only the frequency windows of the radios that were illuminated with white light were clearly visible. The remainder of the instruments on the panel 'glowed' in a shadowy 'pink.' But even the marginal visibility of the instruments stood out in stark contrast to the blackness outside.

The airplane was cruising at eight thousand five hundred as I turned onto the Victor 16 Airway that would take us through the Banning Pass. I looked ahead into the night, and could see a wall of clouds below and nothing but blackness above. As I looked down onto the undercast, the tops of the clouds seemed to be a couple of thousand feet below my altitude and wouldn't be a visibility problem, since I could see the lights of San Bernardino on the horizon ahead. That meant that the undercast was breaking up and scattering. Maybe the weather in the basin wouldn't be too bad after all.

I looked at my watch. It was 7:30 p.m. I was at least another hour from Long Beach. My body ached from sitting all day. I'd been at the controls about seven hours without moving, and every muscle in my body ached. My mind seemed to have slowed down as well. My ability to reason seemed to be in a quiescent state; and calculating numbers or even trying to stay ahead of the airplane was impossible. No wonder the FAA doesn't think long cross country's are good for student pilots. Now I understood why, but too late. My mental processes acted as if I was drugged.

Slowly, very slowly, the realization occurred to me that I wasn't pilot in command anymore. I'd gone from pilot to passenger in my own airplane, a change so subtle it was imperceptible. That should've scared me wide awake, but the long hours of concentration and worry had taken their toll.

I dialed the Ontario Flight Service frequency into the number one radio. I needed to get the current weather for Long Beach some sixty miles west. In my exhaustion I had no idea of my exact position, and it never occurred to me to use the VORs for triangulation. My mind was beyond caring about such life saving actions.

After several calls to Ontario with no answer, I gave up and dialed in Los Angeles Approach Control, and I again confessed my plight.

'Los Angeles Approach, this is Cherokee 8369 November. I'm a PA-28-140, Student Pilot, on a solo VFR flight from Sky Harbor to Long Beach. I think I'm somewhere near Ontario, but I'm not sure. Over.'

'Cherokee 8369 November, Los Angeles Approach. How much fuel do you have on board?'

'69 November has about an hour's flying time. Over.'

'Okay, 69 November,' the voice said calmly. 'Change frequency to 121.5 now.'

Slowly, I set the emergency frequency into the radio. I was so tired that all I wanted to do was to lay down and close my eyes. With someone at the other end of the radio telling me what to do, at least now I won't have to think All I had to do for the rest of the flight was stay awake and follow orders.

'Cherokee 8369 November, are you on frequency?' a new voice asked, tentatively.

'This is 69 November. Over,' I returned cheerfully, glad at last to have reestablished contact with my rescuers.

'69 November, squawk 4133 and Indent. And how many soul's on board?'

'Just one soul onboard,' I replied, as I dialed in the transponder code.

After I pushed the Ident button there was a long pause before the voice returned.

'69 November, radar contact. We're showing you fifty miles northeast of Long Beach, near March Air Force Base. Turn left slightly to a heading of 250. And what's your altitude now?'

'69 November's at eight thousand five hundred.'

'Roger. 69 November, descend and maintain four thousand five hundred.'

'69 November,' I acknowledged, lowering the nose and reducing the throttle by a hundred rpm.

The clouds began to spread wider and wider apart as the airplane traveled further west. And by the time I got down to four thousand five hundred, the entire Los Angeles basin was laid out before me with a trillion lights sparking like jewels. But the endless brilliance couldn't orient my mind as to my position, or give me renewed confidence in my ability to reason. I was just too tired and there were too many lights. I had no sense of direction now and the lights only helped distract me from flying the airplane.

Every few minutes the voice would give me a new heading to fly, redirecting my path towards Long Beach. But try as I might, I couldn't guess at where I was. All I knew, for sure, was that I was in an airplane flying above a sea of lights, somewhere over Los Angeles headed towards the blackness of the horizon that I assumed was the Pacific Ocean.

Among the dazzling melange of light I saw several rotating beacons beckoning me, like diamonds flashing among a bed of sequins. For an instant I thought the beacon off to my right wing was Long Beach, and I was about to tell the voice on the radio that I didn't need his help anymore, but as I looked over the miles of endless geography still to come, I knew it couldn't be Long Beach. It was too far from the ocean. I was still lost, with no direction home.

As in a dream, the airplane continued flying over the vast glow of brilliance that was Los Angeles at night. Nothing happened quickly enough for me now. I wanted the flight to be over. I wanted to be on the ground right now. And the more I wished for closure, the more things occurred in slow motion.

As the minutes passed with no further commands from the voice; I was afraid that he'd forgotten me. And just as I was about to ask if anyone was still there, the voice would suddenly command me to do something, like 'descend and maintain two thousand.'

Now that I was down to two thousand, I had to wake up. Before, at altitude, I could rest and enjoy the ride, but now that I was much closer to the ground meant that I was getting close to Long Beach. I needed to be ready to land the airplane. I needed to be alert, and landing the airplane was one thing the voice couldn't do for me.

'In two miles,' the voice said suddenly, 'I'm going to turn you onto a three mile final for Runway 30 at Long Beach. When I do that, I'll hand you off to the Tower.'

'69 November,' I acknowledged, 'I'll be ready.'

The flight was almost over. I couldn't let up now till I was on the ground. I had to concentrate on what was happening. Ahead and off to my left in the sky, I could see a string of landing lights, in trail of one another, all lined up for Runway 30.

Several minutes later I too was in line, in trail of others about to land. And as I began to ready the airplane for landing, the voice handed me off to the tower and talked to me for the last time.

'69 November, contact Long Beach Tower now, on 119.4.' And then as an after thought added, 'and good luck.' And, with that he was gone.

I quickly dialed in the new frequency for Long Beach and seconds later a new voice called for me.

'Cherokee 8369 November, Long Beach Tower. Are you with me?'

'Roger, Long Beach. 69 November's with ya.'

'Cherokee 69 November, cleared to land, Runway 30, wind calm.'

From two miles out, Runway 30 looks big and friendly. It's twelve thousand feet long and two hundred feet wide, and with its high intensity lighting system, VASI, and taxiway lights all turned up bright, the airport area ahead looked like the friendliest place on earth. The sight of a gigantic field blazing with hundreds of lights of blue and white and red and green, all directing my path to the runway, was just the boost my spirits needed. Just the sight of it gave me confidence that my landing would be assured.

After all of the things that had happened to me during that one flight, the landing itself was 'a piece of cake.' As soon as the wheels touched down, I felt total relief and my entire body went limp.

The flight was over and I was back. I didn't realize just how much tension and pressure I'd been under and how much it had fatigued me, until it was over.

My body felt like a pretzel. I parked the airplane and secured it in the dark, then slowly limped to my car parked near the FBO. Every part of my body hurt.

And the excitement that I'd expected, once I'd flown solo on such a long trip, never occurred; it must've been blocked by the pains in my legs, neck and back.

•

The next time I saw Bill Nielon and I told him about my long solo cross country, he just shook his head.

'I really should've been there to approve it,' he said. 'That's really too long a flight for a student pilot, even in good weather. Ya get too tired flying that much in one day.'

I knew he felt guilty about his absence, and I even forgot to ask him how he did in the Flight Engineer's school, and whether he had passed the 'written.'

•

Captain Richard Graff

It was several weeks after my solo cross country that Bill Nielon returned and told me about the dual night cross country flight every student was required to make for 'night orientation.' Even though he knew that I'd already logged several hours of 'night' during my extended solo flight, he insisted that this requirement was valid, and that it was his policy that a different CFI would fly with me for this flight.

Skeptical as to the value of this requirement as I'd already flown in darkness and survived, I nonetheless readied myself for the flight with a CFI by the name of Richard Graff.

The words **"Capt. Richard Graff'** were embossed in gold leaf on the top of Richard's genuine leather flight case, the kind that airline pilots can be seen lugging through the hallways of airports. By the worn looks of his, one would assume that Richard just got tired of flying 'big iron' and settled down to a quiet life of instructing students. But since Richard was only in his early thirties, that assumption really didn't wash.

Although I'd talked to Richard many times in the office of C&J Air, while scheduling a lesson or reserving an airplane, it wasn't until my training was almost over that I had the occasion to fly with him.

Richard didn't seem to portray the right image of a professional pilot. He was of average height and average weight and bushy brown hair with huge bushy eyebrows. He reminded me of Walter Mathow. But, I always thought his image was more like that of a veterinarian. The way he carried his flight bag always reminded my of a vet's medical kit.

It turns out that 'night currency' was a phenomenon that occurs after a pilot performs three landings during the hours of darkness. Of course, at the time, I wasn't looking forward to any of this night flying stuff, as my landings weren't all that good, even in the daytime. I was sure that having to perform bad landings in the dark wouldn't improve my skill any. When Richard assured me that it wasn't any big deal, I started to worry. I already had experienced his cavalier attitude concerning safety of flight issues during my short solo cross country to Palm Springs and back

On that occasion, I needed the approval of a CFI, and as I recall, Richard was the only one available at the time to endorse my log book. He did so as a matter of formality, having never flown with me before, and having no idea as to my competence. On that occasion his parting advice to me, was, 'if ya run into any trouble, just land somewhere.'

His attitude that nothing in flying was ever a 'big deal' was what kept me constantly on my toes. And I decided that if I was to survive as a student pilot, with CFI's like Richard for teachers, I'd have to hold myself to the highest personal standards of judgment and safety.

Richard and I had a tacit agreement; if I was going to learn good judgment and all that other 'boring stuff' to survive as a flyer, I'd do it on my own. And, as I look back, Richard's methods really forced me to form good habits. Of course, he did it by scaring me and by appearing to be totally incompetent and unaware of the dangers of flight. It was a real 'Baptism by fire;' or was it?

But to be totally honest about it, Bill Nielon had hammered safety and judgment into my flying skills every chance he got; so I did come to Richard with a fair amount of background upon which to draw. But even when I flew with Richard, I was always on guard, since he never bothered reviewing with me 'all that boring stuff.'

Richard's M.O. was like the college professor who says: 'Learn the subject anyway you can; by talking to knowledgeable people; by reading anything you can lay you hands on about it; it all works. The important thing is to learn it.' By this method, I wondered sometimes why I was paying CFIs twenty dollars per hour to learn how to fly?

•

Richard had set up the lesson with me to take place on a Tuesday evening. We would fly from Long Beach down the coast to Gillespe Field in the San Diego area of El Cajun, and then to Hemet, where we would do three takeoffs and landings to a full stop, in the dark; and then return to Long Beach.

The flight seemed simple enough, so I carefully planned it by tracing the route with a red felt tip pen on my Sectional. When I was done, it looked like an isosceles triangle standing on its head, with the Apex at Gillespe. Little did I know how many lessons I would learn during this flight.

I got to the airport about 4:30 in the afternoon. It was the middle of April and daylight would last till after 6:30 p.m. I took my time preflighting the Cherokee 140. It was a nice enough airplane to train in, but it was really under powered for cross country travel. On a good day it would cruise at a speed of one hundred and ten miles per hour in a 'no-wind' condition. Speed was not one of its virtues. The fact that it took a very long time to climb to altitude made it an okay trainer, but not an airplane you'd want to sit in for hours and hours. Although, any airplane, no matter how slow, is better than traveling by car.

With all of its shortcomings, the Cherokee 140 does have one thing going for it–stability. In the air it is rock solid and a pleasure to fly. And by this time I had over fifty hours at its controls and felt very comfortable flying it.

By 5:15 p.m. the airplane and I were ready. And, as if on cue, Richard Graff stepped out of the doorway of C&J Air, flight case in hand, and strolled across the ramp, like a Captain meandering across the tarmac to his 747.

'All set?' he queried, stuffing his flight case in the back seat and deftly sliding into the seat next to me.

'Sure am,' I answered, still sorting out aeronautical charts on my lap board.

'Got a flashlight handy?'

'Two of 'em, right here,' I replied, pulling one of them out of my map case that was wedged between the two front bucket seats, and held it up for him to see.

'Good!' he exclaimed, buckling his seat belt. 'Then we're ready. Let's go.'

And with that, I started the engine, checked the oil pressure, turned on the radios and dialed in ATIS, writing down the information. Glancing over at him, I told him I'd open our VFR flight plan with Flight Service after we were airborne.

'Don't bother,' he shot back, glancing down at his watch. 'I'm with ya, nothin' bad's goin' to happen.'

Hearing that, I thought, oh God, so much for flying safely in the dark.

We took off with the sun low in the sky, and followed the coastline south of Long Beach towards San Diego. Twenty minutes later, as we were passing Laguna Beach, I looked down on the coastal hills and saw their tops now flattened and bare, where developers had already begun bulldozing in anticipation of endless tracts of condominiums.

We flew on in silence, occasionally watching the sun as it sank on the western horizon, almost touching the ocean over our right wing tip. By the time we flew over Mission Bay and Montgomery Field, the sun was fading below the horizon. We turned eastward towards Gillespe, ten miles inland. I'd never been to Gillespe Field and had no idea of its layout. All I knew, was that it was nestled between two ranges of hills. As we flew along, the last rays of the sun were making long shadows across the tops of the hills where the airport was located.

Over the airport from pattern altitude, the shadows made the view downward a confusing series of strobe like flashes–alternate contrasts of light and dark pulsing at one hundred and seventy feet per second. It was like trying to see beyond a picket fence from a moving car.

I watched the show for just a few seconds realizing quickly that it wasn't such a good idea, as 'flutter vertigo' was too easily acquired while flying.

We were number five in the pattern, in trail of four others. The others were so far in front of us that we could only see the number four airplane ahead. I wasn't sure exactly of where I was or what was really going on, as I blindly followed the airplane in front of us. And, after what seemed like a long time, we were cleared to land. I don't know what Richard expected from me in the way of a landing, but this landing was my normal 'abrupt arrival,' that still needed a lot of work. And as we taxied off the active runway, I still wondered why on earth we were landing here?

Since I'd already flown my long cross country that turned into a night flight, I felt that this trip was totally unnecessary, except for the night currency part. After all, I'd already logged a couple of hours of 'night.'

•

The reason we stopped at Gillespe finally became clear as Richard told me he wanted to eat dinner there and wait for total darkness before pushing on with our lesson. So, after parking the airplane and walking over to the restaurant, we discovered that the restaurant was closed. So Richard ate his dinner out of a vending machine, as I preflighted the airplane again, just for practice, and then sat in the airplane waiting for darkness and Richard.

We got underway after Richard returned. The flight from Gillespe to Hemet would have been interesting for me to fly in the daytime. The track of our airplane crossed a series of mountains enroute to the Perris Valley, and since my childhood dream was to someday become a bush pilot in Alaska, it would've been a chance for me to fantasize while winging my way to

Hemet. But now, in the dark, there wasn't much to fantasize about. The only thing I could see was the lighted instrument panel two feet in front of me, now glowing red from the cabin's night vision floodlights.

As we climbed to a cruising altitude of five thousand five hundred in the crisp night air, the cabin took on a chill. I was always more comfortable driving or flying when the temperature was on the cool side, to keep me from getting sleepy. Richard, however, had other ideas. He was getting cold and without discussing it, he pulled the Cabin Heat vent full open, instantly flooding the cabin with uncomfortably warm air. After that, as I sat there half smothering in the heat, I wondered just how much more uncomfortable I could get. I didn't have to wait long, as Richard fumbled in his pockets, took out a cigarette, lit it and blew a thick steam of smoke up at the cabin ceiling.

There he was, comfortable and at ease with the world; and here I was, trying to fly the airplane in the dark, ready to learn whatever he was going to teach me, while trying not to gag on his smoke. The smoke was really irritating my eyes, but I didn't complain, since he was the Master and I was the student.

From Gillespe, Hemet is fifty miles, and in a Cherokee the trip takes forty minutes. In such a short time I wasn't sure that I was going to learn very much at all. But, when you fly with 'Captain Richard,' if you learn nothing else, you'll learn whether or not you ever want to fly with him again.

When preparing for the flight, I had taken great pains to mark my course with a red felt tip pen on the Sectional chart, so it would be easily visible. Above each leg, I printed the compass heading numbers so that they would stand out in the dim light of the cockpit. I had planned the flight down to the last detail, and was now ready to fly my plan.

•

Our climb to cruise altitude engulfed in the night's blackness was over the rolling hills of El Cajon, and as we climbed, I readjusted the cabin floodlights to their brightest intensity. And as the engine roared at full power, Richard lectured me on the pitfalls of night flying, half shouting to be heard over the deafening sound. With so much ambient noise, I wasn't able to pay attention to much else other than maintaining the climb rate speed and heading. His lecture lasted ten minutes, and when he finished, he began to review with me the flight planning I'd done for this trip.

It was just after we leveled off when his inquisition began. Once at altitude, with the power reduced to a cruise setting and the engine leaned, the ambient sound level was low enough that we could talk in a normal tone.

'What's the heading you planned to Hemet?' he asked clinically.

It was one of those questions that all student pilots are asked during the course of a flight, It's really a 'no brainer.' All I had to do was look down at the Sectional in my lap and read off the numbers that I'd printed above the course line. And to make all of this easy for me, I had a small rechargeable flashlight with a red night vision lens in my map case. So as soon as he asked about the heading, I pulled out the flashlight and turned it on, and pointed it at my lap.

That's when it happened; the first of a series of things that I was totally unprepared for. I pointed the flashlight down at the Sectional and looked for the numbers. Nothing! There was nothing visible–no lines, no numbers anywhere in sight.

I held the flashlight closer, thinking that it wasn't bright enough to see them. Still nothing. Finally, I rested the flashlight directly on the map itself, touching the lens to the paper. Again, nothing. I could see the Sectional's depiction of the geographic terrain features fairly well in the faint pinkish light, but no felt tip markings anywhere.

'Sonnofabitch!' I whispered under my breath, totally perplexed.

'I'm still waiting for the heading to Hemet,' Richard asked again, his voice sounding impatient. 'What is it?'

'I don't know. I knew it when I planned the flight, and wrote it down on the Sectional. I swear. I marked the entire flight–all three legs on the map, but now nothing's there.'

Like the sound of a hand grenade, a guffaw exploded next to me, bursting from Richard's mouth and filling the cabin with thick cigarette smoke. As the guffaw changed to a sustained choking cough, he tried to catch his breath and laugh at the same time. And the sounds of laughing and coughing lasted almost a half a minute, trailing off finally into an inaudible snicker.

'God, I love this part of night orientation,' he began. 'Don't worry about it,' he laughed again. 'Everybody makes the same mistake. That's why we take ya up at night. Ya have to learn certain tricks to stay alive while flying at night, and one of 'em is, ya never, ever use the color red to mark a navigational chart or anything else ya want to read in the dark. The red night vision floodlights make everything that's colored red disappear.'

'Aaahhh,' I groaned, suddenly understanding.

'Now I could've told ya all this yesterday,' he snickered, taking a deep drag on his cigarette. 'But, I think showing ya's a lot better than telling ya, cause now ya'll remember it longer.'

'Gee, I never knew that,' I said, humbly.

'Red lines, under red light, disappear. And now that you've experienced the drama of it, ya'll never forget it, as long as ya live. I'll guarantee it.'

I suddenly saw 'Captain Richard' in a totally diffferent way. Before, I only thought of him as cavalier and a buffoon; but tonight, I experienced his cunning and resourceful teaching methods. And the biggest shock of all was that he'd actually taught me something. Not only that, but he'd taught me something in such a dramatic way that I knew I'd never forget it.

'In certain situations, using red markings could be a killer,' he continued. 'For instance, on a night flight,' he gestured with his left hand holding the glowing cigarette. 'If ya really needed to know a heading to keep from hitting a mountain peak, ya'd never use red to mark with,' he said, looking over at me.

'I know I'll never forget that trick,' I said, nodding. 'But it doesn't help me tonight to find Hemet. I'm not sure I remember what the real heading is.'

'That's easy,' Richard replied, as he glanced down at his watch. 'From here, the heading to Hemet is 346 degrees, and it's about forty miles and should take us about another twenty four minutes.'

'Something tells me you've flown this route before,' I grinned.

'Ya, a couple times,' he snickered.

•

The evening was as black as pitch as we flew northward. The sky was overcast above us, so no moon or stars were visible. The route northward to Hemet paralleled Interstate 15 from Gillespe, winding past Temecula and Escondido before finally aiming more or less towards Hemet. Except for Escondido, there were only a few pin points of light from the outskirts of Carlsbad and Oceanside some thirty miles off our left wing tip

— certainly not enough light to help a pilot separate earth from sky.

We flew on in silence for several minutes, the two of us sitting there in the dark, suspended in the sky, a mile above the earth with no visual cues and no sense of motion. We could've been sitting in a flight simulator on the ground somewhere, listening to simulated engine noise and moving the controls to keep the instruments in harmony. But, gradually, off to our left at our ten o'clock position, I saw the dim silhouette of the two peaks of the ridge line of the Santa Ana Mountains moving ever so slightly towards us. We were some distance away and off to one side, so that crossing them was never a consideration. It was the local reporting point for Orange County Airport, 'Saddle Back.' It's outline stood in faint relief against the dim city shine of Los Angeles.

The faint outline of the mountain ridge made me wonder if I would know, when flying at night, whether I was high enough to clear a mountain ridge. Richard must have been reading my mind. He immediately began lecturing on how a pilot could determine whether his airplane was of sufficient altitude to clear a mountain in the dark. And after his tutorial, it turned out that if one is able to see lights from beyond the summit of a mountain, then one's altitude is sufficient to clear it.

That was the second lesson that I'd learned from Richard tonight. My respect for him as a teacher was growing. I guess I'd never considered his teaching skills before, since his comic bravado and macho nonchalance always got in the way.

We started letting down about fifteen miles from Hemet, probably somewhere around the town of Perris Valley. And by the time the lights of Hemet were visible, we were just flying over the airport, checking the tetrahedron as to wind direction. We crossed the field at four thousand and surprisingly, the tetrahedron was clearly visible from this altitude, in the center of a ring of lights.

A few minutes later, we were on the ground.

●

Originally, Richard directed me to plan a flight from Long Beach to Gillespe to Hemet, and then back to Long Beach. But now that we were on the ground at Hemet, he had another surprise for me.

'Ya know, Bill,' he said slowly. 'Part of the reason for a supervised night flight is to get ya 'current for night.'

'What does that mean, exactly?' I asked, as we sat on a taxiway with the engine idling.

'To be current at night is almost like being current for day,' he continued. 'Ya have to do three landings at night to a full stop. And if ya do them every ninety calendar days, then you're always current for day and night.'

'So it's a little different from being current for days,' I replied. 'Since for days, ya don't have to do full stop landings, just 'touch and goes.'

'That's right,' he said, lighting another cigarette. 'For days, ya just have to do three touch and goes.'

'So what are we going to do now?'

'Well I was goin' to have ya get night current here, but I changed my mind. This place gets too busy at night. I think we'd be better off at Corona,' he answered, pointing over his right shoulder. 'Let's take off and go over there. It's uncontrolled and there won't be anybody there to bother us.'

So with that, we took off and flew to Corona. Since I hadn't planned it, I had no idea which direction it was. But as we climbed out on takeoff and before I could ask, he yelled over the noise of the engine, 'The heading's 285 degrees and fifteen miles.'

I nodded and banked the airplane to the right, picking up the heading from the faint pink numbers of the directional gyro.

'Level off at two thousand,' he yelled. 'We don't want to get too much altitude.'

Within twelve minutes we were in the pattern at Corona, and even though the airfield there was small, uncontrolled, and three miles out of town, it had Pilot Controlled Lighting. I was impressed, as even some larger city airports still didn't have that capability. So keying the mike on the Unicom frequency seven times within five seconds, turned on all of the lights of the runway and taxiway. At Corona only the windsock stayed illuminated twenty four hours a day.

Actually, getting there wasn't a problem, I just flew towards the lights of Corona. It's urban sprawl generated all the 'city shine' I needed. I'd only flown there one time prior to this, when Bill thought I needed to experience some 'cross wind' pattern work that turned out to be a total waste of time; I just couldn't do anything right that day. The one thing I remembered about Corona was that it had low hanging power lines stretching across the approach end of Runway 25. But to keep pilots from hitting the lines on final approach, six large plastic balls the size of basketballs and painted 'day glow orange' were attached to the wires every twenty feet.

But, landing there in the dark and passing the wires on final at eighty miles per hour didn't allow me time to see either the wires or the balls. I was too busy watching the runway that was lit up like an aircraft carrier. I focused on the first third of its length where I intended to touch down. The runway was only thirty two hundred feet long and I didn't want to land long, and run out of pavement or airfield.

•

I was never really good at landing Cherokees. There was something about their 'Hershey bar' shaped wings that never let me land smoothly. Years later, when I began to fly high-winged Skyhawks, I didn't seem to have the same trouble. I think the reason I always had trouble landing low wing airplanes was because no instructor really took the time to teach me in detail the technique required for setting up a good landing.

Years later, when I bought my Pacer, Bill Hare spent twenty five hours teaching me in great detail each step in the landing technique. But during my training for the Private License, everybody just kept telling me not to worry, that I'd get the hang of it, eventually.

Anyway, I knew that my night landings were going to be pretty much like 'controlled crashes,' since my daytime landings were not improving much beyond that stage, turning into 'abrupt arrivals.' But, having experienced only one other night landing prior to this flight, and with nobody able to teach me the visual cues for good landing technique, I wasn't sure what to expect; I just knew that it wouldn't be pleasant to endure.

•

Since Richard always gave the impression that nothing about flying ever bothered him, I was game if he was. I knew that if it'd been Bill Nielon, he'd critique my every move–which would've only made me nervous. And as it turned out, it would be several years later before I'd discover on my own what the correct visual cues were.

As we landed the first time and rolled out, I took most of the runway, and Richard's only comment was to relax and try to do everything smoothly. The problem was that I didn't know what I was suppose to do to be smooth about.

'After all,' Richard said, looking over at me. 'Who's going to be watching if ya make a mistake?'

I nodded.

'Besides, ya gotta practice sometime, why not now?'

So I practiced and did three circuits, each to a full stop. The initial landing really was an 'arrival,' like an Navy fighter plane slamming onto the deck of an aircraft carrier.

The first landing of the three circuits for 'night currency' was better than the 'arrival,' but I had trouble judging the airplane's height above the runway when I got to the 'round out,' eight feet in the air. What I would discover years later, was, at that point, to look at the far end of the runway to better determine the airplane's altitude above the ground. Once I learned that, I could tell within a foot or so where the wheels were.

The second landing was similar to the first but, undaunted, Richard quietly sat staring out the side window at the darkness, puffing on his cigarette occasionally grunting, 'Just relax. It'll come with practice.' And then he'd go back to staring out the window again.

The third landing of the three wasn't much better than the other two, but with the added feature that we bounced a couple of times, finally getting the airplane on the ground before using up the entire runway.

By that time, even Richard had had enough, and decided that we better return to Long Beach, and call it a night.

And so, as we climbed out of Corona to two thousand on another of Richard's headings, he casually turned toward me and said, "Not too bad for your first time at night. It takes a little practice to get the hang of it, that's all.'

The flight back to Long Beach was just twenty five minutes on a heading of 258 degrees in weather that was 'severe clear.'

And as we flew along, I wondered if I would ever learn to fly an airplane as good as Richard.

•

That night flight was the last time I ever saw 'Captain' Richard Graff, although I continued to rent airplanes and fly out of C&J Air for the next two years.

I guess he was just never around when I was. And then, one time, someone thought that he'd left the area to take a flying job with a small air taxi outfit in northern California. And, although I'd ask about him from time to time, no one really seemed to know for sure where he was.

I kind of miss him. And every now and then, I still think about him.

Chapter Three
If Today's Sunday, Then this Must Be Oceano

A Gathering of Sunday Pilots

What's a fly-in? Where's it held? And, what goes on there? These are all questions to be answered for someone that's never attended one of these happenings.

Fly-ins are meetings of flyers, all of whom fly the same kind of make and model airplane. They belong to clubs called 'type clubs.' And fly-ins are meetings where members of a type club fly into a specific airport and meet for a half a day to check out each other's airplanes, eat lunch and do some hangar flying, before departing. Fly-ins are to pilots as car rallies are to sports car enthusiasts. It's a chance to get together and show off a little. Anyone can host one. All one has to do is to advertise that, on a certain day at a certain airport, there will be a fly-in sponsored by whatever club, association, or group. It usually starts at about mid morning on a Saturday or Sunday and lasts until late afternoon. Not only does it give the people a chance to get together to socialize, but it gives them a reason to fly somewhere in their airplane.

It's a chance for owners to trade maintenance secrets, to share to repair problems, and parts' outlets, where to find parts, and to see how the same airplane can be modified in numerous ways. And after several hours of checking out each other's airplanes, the group usually meanders off to lunch at an airport coffee shop or restaurant within walking distance. And, after a leisurely bite to eat and visit with members that live great distances apart, they adjourn to their airplanes and fly home. Thus, not only do they enjoy several hours of great flying, each way, but they spend quality time with other aviators.

The reason fly-ins flourish is because it offers pilots a destination for a weekend flying trip; and it assures the attendees a pleasant day of camaraderie. At regional fly-ins, where several states worth of airplanes gather, the program agenda might be several days in length, with flying games to be played, such as spot landing contests. Spot landings compete pilot against pilot to see who can control their airplane's so as to touch the airplane's wheels down exactly on a designated 'spot,' (usually the runway threshold or runway numbers). It takes skill and practice to land exactly where you want to.

Another game that's always fun at fly-ins is the bombing contests. Bombing contests also pit one pilot against another for bombing accuracy; five pound-flour bags are dropped out of their airplanes at an altitude of five hundred feet, usually at a painted 'bulls eye' on the airport's infield.

Then, finally, there's the always popular poker run, where each airplane takes off and lands at each of four other designated airfields, picking up one playing card at each field before returning. When all the airplanes return from a poker run, picking up the last card at the initial airfield, the pilot with the best five card poker hand wins.

And, finally, no fly-in is complete without a raffle that gives away a dozen or so door prizes.

My flying buddy Les and I spent many a weekend over the years at various remote airfields at fly-ins. The following describes one such gathering, and illustrates why it's always much more fun to attend a fly-in with all its festivities than to host one with all of the work.

•

I'd bought my first airplane in 1983, after renting airplanes for ten years. It was a PA-20, Piper Pacer, built in 1950. And, since it was a high wing, tube and fabric airframe with a 135 horsepowerd engine, it was quite different from the 'spam cans' I'd rented from various FBOs over the years. It looked like it was born of a different age, and it was. It wasn't show quality but I didn't care. It was a classic, even though it's only radio was a 'stone age' one using tubes, in the age of semi-conductors. I didn't care. I bought and paid for it. It was special to me. Especially after flying the boring production models that FBOs had on their rental lines: either Piper or Cessna, with a Grumman American thrown in for good measure.

After flying my Pacer for several years, my friend, Les, decided that he too, could afford to buy an airplane. Of course, since both of us were 'working stiffs' we couldn't afford anything new or expensive (that translates to anything 'high performance'). What we each ended up with was dictated directly by what our bank accounts could afford.

As it turned out for me, for the price of a modest used car I was able to buy the Pacer that had been an Alaskan bush plane for the preceding 33 years. The Pacer was the last of a series of tail-wheeled airplanes built by the Piper Aircraft Company.

And, I bought it for two reasons: first, because after ten years of renting airplanes, I was really tired of paying rental prices; and second because a classic tail dragger from Alaska looked like it would be a joy to fly.

For the first ten years after I earned my pilot's license, I'd flown rented airplanes — the same two or three makes, the only ones for rent on any FBO's line.

The rental choices were either two or four place Cessnas, or two or four seat Piper Cherokees, with a Grumman Tigers or Travelers available now and then. And, after a while, flying the same old airplanes was like renting the same old car. Even Hertz had more than just two makes to choose from. Not only did the same old hardware make for boring flying, but it hardly justified the dollar and change per minute rental cost. For a buck a minute, it just wasn't a buck a minute worth of fun.

The thought of renting the same old 'spam cans' was depressing, and for months I didn't know what to do about it. I was frustrated to the point that I almost quit flying, altogether. I moped around in a 'funk,' because I used to love flying, but now it wasn't enjoyable anymore.

I spent months trying to sort things out, trying to understand to where the fun of flying had disappeared. Finally, I concluded that rental airplanes were boring and a dead end. What I needed if I was going to continue to fly was a major change, like a completely different airplane something novel perhaps a classic or antique.

As for unique airplanes, for me, the low winged all aluminum Ercoupe with its two-place side by side seating and its bubble canopy and twin vertical Tails seemed to be the answer. It was a vintage airplane, designed in the thirties, but because of its advanced technology, by today's standards it's considered a state-of-the-art design, more than many current production airplanes.

Because of its small size and cartoon-looking shape, the

first time I saw one it appealed to me, the way a VW Bug appealed to me in the fifties. I saw my first Ercoupe when I was a kid, at the Cleveland Municipal Airport. It was parked on the ramp almost under the wing of a DC-4. And by comparison, it looked like a toy, like something out of Walt Disney.

So, for several months I researched the evolution of the Ercoupe with all of its models. I even went on a crusade of sorts, checking out various 'Coups for sale, while looking for the 'right' one. The search extended from Los Angeles to Livermore to Casa Grande to San Diego. I visited every airfield large and small, controlled and uncontrolled that was published on the California and Arizona Sectionals.

I finally found several that I liked and was interested in flying. But, just as soon as I got serious about flying them, I discovered that I had a major problem: either the airplane was too small for my body, or my body was too big for it. Either way, I didn't fit. And no matter how hard I tried, I couldn't fit my body inside and still close the canopy. Even with the canopy open, my knees couldn't clear the instrument panel–and of course, in 1937, the seats didn't adjust; you sat where the seat was, no matter how long or short your legs were. Even if my knees cleared the panel, my thighs kept the control wheel from turning. And my head was so tall that the canopy couldn't be closed. There was just no way I was ever going to own a 'Coupe.'

I couldn't believe it. My favorite vintage airplane and I didn't fit. So much for fantasies. It turned out that the problem with the 'Coupes side by side seating was that the interior was only thirty nine inches wide. It must have been sized for children or midgets. So if two 'normal' sized people–whatever that means–tried to fly in one, they'd takeoff as strangers, but land as 'intimate' friends. Coziness was one thing, but for me, it was uncomfortably cramped.

So I resigned myself to owning whatever vintage airplane I

happened to fit. Like Cinderella, I went on a crusade of trying on airplanes for size.

But even though I could never own one, my very favorite model was the 415-C, with its eighty five horse Continental engine and 'two-control' flight control system that doesn't require rudder pedals. The rudders and the ailerons are interconnected, like modern day high performance airplanes.

Most contemporary pilots don't know what they're talking about when they assume that since the 415-C has no rudder pedals, it also has no rudders. Actually, the airplane has twin tails and was very advanced for the late thirties. And its interconnected control system that makes every turn and bank a coordinated one, with the rudders automatically deflecting just the right amount when the control wheel is moved. This makes it as modern as a Cessna 210 or a Beechcraft Bonanza whose controls systems are like that of the Ercoupe.

Although not being able to own a 'Coupe was very disappointing to me, I turned my attention to other classic airplanes; and there weren't many. In fact, there weren't any! The closest I could come to airplanes that were remotely different from what I'd been renting were vintage tail draggers.

I really hadn't considered tail wheeled airplanes, as they're not flown much anymore, and getting instruction in them is almost non-existent. Getting instruction in them is very important, especially if you don't have previous tail wheel time, and, of course, I didn't. Insurance companies require an appropriate amount of recent dual instruction in tail wheeled airplanes before offering liability insurance, because when landing, these planes have a tendency to ground loop (doing a cartwheel from one wing tip to another), thus destroying the wings and tail in the crash. Quite a contrast to the modern airplanes with nose wheels.

Despite all of the hurdles, I guess it was my destiny to own a tail dragger, as one suddenly caught my eye as I was driving through my local airport on a Saturday morning. It hadn't been there the day before and surprised me. Not only that, but there was a 'For Sale' sign taped to it's windshield. I quickly stopped the car and ran over to the sign. It proclaimed that the airplane was a 1950 Piper Pacer and gave a telephone number to call.

As I approached its high wing I had to and duck down to clear it. I leaned my face against the cabin window and cupped my hands to shade the light so I could see inside.

The size of the cabin seemed to be equivalent to that of an Ercoupe, but there seemed to be more head room. The anachronistic control wheels and instruments gave the cabin a nostalgic feel. For some inexplicable reason, at that moment, the airplane's old time 'look' suddenly appealed to me–it was so different. The seats were vinyl and needed reupholstering. The headliner was new. I think the sight of it took me back to memories of my childhood, when I had seen similar airplanes when they were new.

Except for an electric Turn and Bank indicator, there were no other gyro instruments. It was strictly a VFR airplane. The vintage 'oval' shaped control wheels gave the final touch of antique authenticity to the cabin's ambiance. For avionics, the only thing I saw was an old Escort I combination navigation and communication radio with a built in VOR, The problem I had with a vintage vacuum tube transceiver was it was heavy; and it could either be used as a navigation receiver or a communications transceiver, but not both at the same time. It was 1950 technology for sure, just like the airplane.

I think what really got me excited about the airplane was its honesty; what you saw was what you got. It just seemed like a basic airplane, with simple and reliable systems. Although it was tube and fabric in the age of metal, it did have four seats.

I copied down the seller's telephone number from the sign

and quickly drove home. Once home, I phoned the Pacer's owner. The owner turned out to be the retired Manager of the FAA General Aviation District Office in Long Beach. His name was Frank Allen and he'd been with the FAA since 1936, retiring in 1975. When he started with the Agency, it was known as the Civil Aeronautics Board, under the jurisdiction of the Department of Commerce. And, to qualify for a job with the CAB, one had to hold an A & P license, which he did.

'What's wrong with the airplane?' I asked. 'Why're ya selling it?'

'There's nothing wrong with it, at all,' he said. 'I just finished rebuilding it from a basket case. It took me over two years.'

He paused for a moment and then continued. 'In retirement, I always wanted the fun of using my A&P license, but I made the mistake of buying a second airplane project before I finished the first one, the Pacer. And now half my garage is occupied with the second airplane, and my wife's mad at me, cause we can't park our cars there.

'So why're ya selling it?' I repeated.

'Well, as soon as I was finished rebuilding it, I was going to have fun just flying it. But now that I've got two airplanes and only one tie down spot at the airport, I've got to sell the Pacer to make room to park the second airplane.'

'Ya mean ya wouldn't be selling it if ya hadn't bought the second airplane?'

'Right! I never planned on rebuilding two airplanes. But now I've got to finish the second one and move it out of the garage or my wife said she'd divorce me for sure.'

It turned out that he'd started rebuilding the Pacer and by the time he'd almost finished it, someone made him an offer he couldn't refuse for another 'basket case.' He said he'd planned on looking around for another project after the Pacer, but he didn't expect to find one so soon, and with a price that he couldn't pass up.

'So, do I understand then, that the tie down spot doesn't go with the airplane?'

'That's right,' he answered. 'If I give it up then I won't have anywhere to put the second airplane.'

'Well, fortunately,' I said, 'I already have a tie down at Hawthorne Airport that I've been paying rent on for the last two months. I anticipated buying an airplane. And, at the moment, there's a five year waiting list on the field here at Torrance for tie downs. And there's at least a two to three year waiting list at most other airports in Southern California.'

'That's really something, isn't it,' he said, shaking his head.

I told him about my size problem with the Ercoupe and my crusade of trying to find an airplane that fit me.

'Tell you what, meet me at the airplane this afternoon and see what ya think.'

So that afternoon I would play Cinderella again, and try on another airplane.

•

When I met Frank Allen he reminded me of Charley Ruggles, the actor. He was the perfect grandfatherly type from Central Casting. A distinguished looking man in his seventies, with thinning white hair, a ruddy complexion, and a pencil thin white mustache. And like Ruggles, he had a soft-spoken voice that seemed to go well with his quick wit.

I told him that I'd been looking for a fun airplane, but most of the vintage models were too cramped for me. So, the first thing he did when I showed up was to open the cabin door so I could try the airplane on for size.

When I climbed into the cabin, the two side by side bucket seats in front were as cozy as those of the Ercoupe, but the amazing thing was that the leg room was really more than adequate, and my head cleared the headliner by several inches.

The fact that I could actually fit comfortably inside of it, impressed me. And after he started the engine and I sat there listening to the purr of the Lycoming 0-290-D2 135 idling at six hundred rpm, I just knew I had to buy it. As I sat there watching the antiquated engine instruments come to life, with their overly large indicator dials and needles, everything just felt comfortable, a feeling that this was my airplane. I wrote him a check on the spot. And when it was done, I realized that I'd bought an airplane without ever flying it.

The one thing I did do before writing the check was to carefully study the ship's papers and logs. Then I gave the airframe a thorough walk-a-round preflight inspection.

When the inspection was over, nothing obvious came to view regarding things loose, hanging or dripping. There didn't seem to be anything that appeared to need fixing; and even if there had, I would've taken care of it myself. I really wanted this airplane.

The engine and airframe log books told me that all of the Airworthiness Directives had been complied with, and that it had a rather low time engine with only three hundred hours of use, since it was last overhauled. The airframe had thirty four hundred hours on it, and had spent its whole life in Seward Alaska on floats. Then, five years ago, it was disassembled and stored away until Frank bought the pieces and put them back together again.

The airplane had been delivered new from the factory in 1950, to Seward, Alaska, where it had operated as a bush plane, hauling fisherman and supplies to and from Anchorage.

As I finished reading the various documents, I realized that

I'd finally found a vintage classic that was nostalgic enough for me to get excited about, and would be more fun than flying modern Pipers and Cessnas.

•

During the mid-eighties, the parking situation for airplanes at all of the upscale airports in the Los Angeles basin was grim. There were waiting lists for tie down parking that were three and four years long; and there were even longer waiting lists for hangars. It wasn't because every parking space was filled. It was because no more of the available tie down spaces were allowed to be filled because of the local homeowner associations. Residential areas of urban sprawl that surrounded each airport were now a powerful lobbying force. And, the number of airplanes on each field was now limited because of new noise requirements.

For instance, Torrance Airport could accommodate over 2,000 airplanes tied down on the ramps, but was now restricted to just 800. The homeowners associations had won. Nobody was listening to the pilots. And so, airplane owners resigned themselves to population restrictions at all Southern California airports.

Noise abatement was a very big deal to all of the homeowners associations. So, at most airports, even though only three quarters of the parking spaces were filled, waiting lists prevented additional airplanes from coming aboard until vacancies occurred. The object was to 'steady state' the airplane population, and it was working.

Several months before I actually bought an airplane, I began looking for an airport that had a tie down vacancy, so that when I bought an airplane I'd have some place to park it. So, I was resigned to the fact that unless I found a parking space, I couldn't buy an airplane. So I began visiting airports to see what the chances were of getting a parking space during my lifetime.

Two years earlier, on a lark, I'd completed an application for a tie down at Torrance Airport, the nearest airport to my house, only five miles away. I had no idea at the time if and when I would ever buy an airplane, but I figured, what the hell—maybe someday I might and it would be nice if I could park it near my house. The people in the Operations Office told me that they had a five year waiting list, but would be happy to add my name to it. So, I filled out an information card that they filed.

'Good luck,' they yelled behind me, as I left the office.

As things turned out, seven years would go by before my name would come up on Torrance's five year waiting list for a tie down space.

All in all, I visited seven airports inquiring about renting a tie down space, with no luck anywhere. But destiny intervened in the strangest way. It was the week when the automatic clothes washer broke down in our house, and forced me on a Tuesday night, to visit the only all night Laundromat in the area, some six miles away. So at 9 p.m., somewhere in Redondo Beach, in a store front Laundromat, I sat alone, surrounded by two dozen washing machines and a dozen large dryers. But, after an half hour, a woman walked in carrying a laundry basket of clothes and proceeded to do her wash.

The woman turned out to be Joyce Jones, a CFII, whose fifteen week Instrument ground school I'd just completed at the local high school. During her evening classes, she and I had become friends, and while our clothes were washing we had a nice 'hangar flying' session. She had spent World War II as a WASP ferry pilot, flying everything from P-38 fighters to B-17 bombers. I think she told me that by the time she was done she had almost four thousand hours of 'tail wheel time' Outside of the P-38 and Bell Aircobra, most all other fighters and bombers were tail draggers.

During our chat I told her my plight. I'd finally saved enough money to buy an airplane, but couldn't find anywhere to park it.

'Do ya know where Rose Aviation is at Hawthorne Airport?' she asked, sympathetic to my problem.

'Ya,' I answered. 'I rented an airplane once from Wright's Flying Service, just down the way from Rose.'

'Good! Cause what ya don't know is, I work at Rose and I know there will be two tie downs opening there, this week; in fact, they're the only two to open for the next two years,' she smiled.

It turned out that the two pilot's were retiring from the area and moving out of state. Otherwise, the waiting list was still three years long.

'Since you're a friend and former student of mine, I could make sure that one of those two is yours; but yaw'd have to come over tomorrow checkbook in hand.'

And so, that's how I came to rent a parking spot at an airport, before I ever owned an airplane. And if I hadn't taken the spot right then, buying an airplane would've been out of the question.

And it was only after I'd bought the Pacer and had Frank Allen deliver it to Hawthorne Airport, to my new parking spot, that I began worrying about where I was going to find someone to check me out in it. But one worry at a time.

•

And so I began another crusade in a twenty mile radius, hunting for a CFI that was current in tail draggers. I phoned

FBOs at Van Nuys, Burbank, Fullerton, Long Beach, and Torrance. Nowhere was there an instructor that could check me out in my newly acquired Pacer. I even went as far away as Camarillo, some fifty miles as the crow flies, but no luck, either. It turned out that most CFI's were on the young side, in their twenties, and unless they'd taken an aerobatics course in a Decathlon, they'd probably never seen a tail dragger before, let alone be current in one.

Discouraged, I complained to Chris Miller, one of the owners of Security Aviation, an FBO next to Rose, at Hawthorne. Since I was out of ideas I thought someone there might have an answer for me. And, as luck would have it, they did.

'Ya came to the right place,' he said. 'There's someone on our staff right here, who can check you out,' he smiled. 'His name's Bill Hare and he's current in tail draggers. Hell, he owns his own Pitts S-2A and gives aerobatic lessons in it. I guess he'd be current,' he smiled.

It turned out that Bill Hare was a CFII who worked full time as a swing shift maintenance supervisor for Western Air Lines at LAX, but six days a week, until 2 p.m. every workday, he was a flight instructor at Security Aviation. And after all of my hunting, he was the only instructor I could find in the entire Los Angeles basin that was current in tail-wheel airplanes. And now that I'd found him, there turned out to be just one other problem–his availability. Chris told me that his work schedule at Security, was Tuesday through Sunday; with Monday his only day off. But, he warned he's booked up for months in advance.

Undaunted, I looked him up. Now at least I'd found someone that could help me and was qualified; it was just a matter of when.

•

Bill Hare was in his late forties, of slight build, a hundred and fifty pounds and about five feet ten. His hair was thick and prematurely gray. He was easy to talk to and a very quick thinker. I introduced myself to him and told him about my plight.

It's hard for me to convey just how gracious and condescending he was to take time out of his busy schedule to talk to a total stranger. He acted really concerned about helping me, as if I was a twenty thousand hour airline captain. It didn't seem to matter to Bill who the person was. If you asked for his advise or help, he'd give it. And at that moment I really needed his help and was impressed with his manner.

'I've got a half an hour before my next student, where's your airplane?'

'Down the ramp a ways.'

Let's go look at it,' he said, starting off at a brisk walk towards the ramp.

•

As we walked, he told me about himself; that he'd been flying since he was in high school, back in Indiana. His Dad had a farm and was a pilot with his own airplane and used to take Bill along with him on trips. Bill pretty much learned to fly by just riding along with his Dad and watching. He said he'd been flying ever since. He said he'd probably flown every type and model of general aviation airplane made, since the J-3 Cub. And now he owned a Pitts Special that he used for teaching aerobatics. It turned out that his airplane was in a hangar within a hundred feet of my new tie down.

When we got to my Pacer, I unlocked the cabin door and Bill gracefully slid into the pilot's seat. He just grabbed a diagonal cross brace on the inside of the windshield with his right hand, and with his right foot on the step, pulled up and slid inside and over in one fluid motion; like a ballerina. A really slick move. It was obvious he was accustomed to vintage airplanes.

He sat in the Pacer for fifteen minutes, taking stock of the airplane's instrumentation and concluded his inspection by checking out the toe brakes on the pilot's side; there weren't any brakes for the copilot.

'Does the fact that there's no brakes on the right side bother ya?' I asked, tentatively.

'Naw. Brakes won't help ya recover from a developing ground loop. Ya just have to be real fast on the rudders.'

When he was done, I asked him if he thought he'd have time to take me on as a student, so I could get checked out in my newly acquired airplane. I told him that AVEMCO, the insurance company that covered the Pacer's liability, stipulated that I take fifteen hours of dual, prior to solo flight, or the insurance was void, since I didn't have any previous tail-wheel time.

'Sure. Be glad to. Just go into the office over there and write yourself into the master schedule for Saturday and Sunday mornings from 8 a.m. till noon, for about five weekends in a row, and then after that, we'll see how it goes.'

And that was that. Of course, it was almost a month later before I could start my weekend training with Bill because of his schedule, but I didn't care. At last I was going to put some fun back in my flying. And besides, I just felt lucky to be able to find training time on Bill's schedule, period.

•

So, finally I began my training for the checkout in the Pacer. And it became a marathon every Saturday and Sunday. It ended up taking more than three months. It wasn't that I was a slow learner, but during the landing phase of flight, I just couldn't feel the subtle sinking motion of the airplane, after it rotated to level. It took my 'bottom' about twenty five hours of dual, before it was sensitive enough to feel what was happening to the airplane. For three months worth of Saturday and Sunday mornings, Bill patiently rode around the pattern with me, explaining what the airplane was doing every minute of every circuit.

As a teacher, he had the patience of Job, even after I was getting impatient with myself. But Bill's teaching technique didn't allow student's to become intimidated. And I remember one Sunday morning, in particular, when he patiently rode the pattern at Hawthorne Airport for three and a half hours, while I completed twenty nine landings to a full stop and taxi back — always 'three-pointers.' It was tedious, but he never complained, and he always tried to keep me from giving up when I thought I'd never 'get there.'

Finally, after twenty five hours of dual, I suddenly felt comfortable and ready to fly my own airplane. At last I'd gotten enough instruction to really understand the aerodynamics throughout all modes of flight. And once I mastered the Pacer, it really did put the fun back in flying for me. It felt like the beginning of a new chapter for me, and for the next six years and four hundred flying hours, I was truly a happy camper, and the Pacer really was an absolute joy to fly.

•

After I bought the Pacer, Les decided that he too could afford to buy a modest airplane. He'd already been a 20-year renter of airplanes. After renting year after year from FBOs just to go flying for an hour or so, it doesn't take long before a pilot wishes he owned his own airplane. It's not an ego thing, just a matter of convenience — a 'known quantity.' With a rental, you never know what to expect: whether the airplane's condition will be ratty, or the equipment isn't the preferred mix that an individual flyer wants for instruments and avionics. And every flight can turn into an adventure, if the equipment is not well maintained or not rented and in use regularly.

Les bought a Mooney Cadet from a man in Fort Meyers, Florida. He flew there commercially to check it out. And, if it looked okay, he'd then buy it and ferry it back to Los Angeles. He did this during the worst possible time of the year; during the Spring season of thunderstorms along the Gulf. His exciting ferry flight (which almost killed him) is documented in Chapter 4.

Anyway, once Les had his airplane, he and I would fly to a 'type club' fly-in at least once a month. We both joined and supported the EOC or Ercoups Owners Club, one of the largest and most active clubs in Southern California. The EOC was a national club with a large membership and well-organized, with regional chapters and a really good newsletter. Because of its size and active membership, they held a fly-in every month. Their monthly fly-in always took us somewhere that neither Les nor I had ever flown before. The fly-ins were always scheduled on a Sunday, so pilots could take care of their 'honey do's' the day before and were freed up to do some serious recreating the next day.

Their fly-ins were fun, and the people were warm. Of

course, Les' Mooney Cadet, an offshoot of the Ercoupe design, was a legitimate 'Coupe relative, and was welcomed by the club; whereas my Pacer was always thought of by the members as a kind of outcast and a 'tag-a-long groupie' that was tolerated.

So, it was no real surprise to me, when Les called me at work one day and asked if I'd like to go with him to an EOC fly-in at Oceano County airport the following Sunday.

'Where's Oceano? I've never heard of it,' I asked, puzzled.

'It's just a small airfield south of San Luis Obispo, near Pismo Beach. It's on the charts.'

'Well, I'm game if you are,' I replied, getting the Los Angeles Sectional out of my desk and unfolding it to its north side.

I ran my finger up the map's depiction of the coastline from Santa Barbara to San Luis Obispo. Just before San Luis, where the land juts out in a nub, there it was. A magenta symbol for an uncontrolled airfield with fuel and the word 'Oceano.'

'Hey, I found it!' I yelled into the phone, surprised. 'What time do ya want to leave?'

'We'd better leave about 7 a.m. to beat the fog,' he remarked, referring to the sudden ground fog that develops at Hawthorne Airport at about 7:30 a.m. each morning in the summer. Once the fog develops, it doesn't burn off until after 10 a.m., when the field finally clears to VFR. And getting out of Hawthorne VFR once the fog develops is rather 'iffy,' even with a Special VFR Departure. It all depends on how thick the fog gets. And since neither of our airplanes were certified for IFR flight, departing on an instrument flight plan wasn't an option.

Since Les bought his airplane he attended every fly-in he could find, not only to show it off, since it was one of the few remaining Mooney Cadets still flying anywhere, but because after twenty years of renting, it was a novel feeling for him to go flying whenever he pleased, without phoning in a reservation to the FBO like a renter.

During the time that Les was acquiring his airplane, I was still refurbishing my Pacer with Walt Wasowski, a retired American Airlines mechanic. Walt and I worked on the Pacer every weekend, rebuilding all of its systems. When we were done, the only thing left to do was to paint it. But, I decided not to make it look too good for vandals and thieves, so I left it with its weathered paint scheme, as a kind of 'urban camouflage.' That way, with a ratty exterior look, the curious might not think it contained anything of value to steal. But inside the Pacer's cabin, it was another story. It contained new leather seats, a really good panel that included not only the 'big six' using a venturi system for a vacuum, but a DME as well, something unheard of in a vintage airplane.

●

It was always fun for me to go to the fly-ins in the Pacer, especially to the Ercoupe ones, as tail draggers like the Pacer require a higher level of flying skill to land than 'nose-wheel' airplanes, like the 'Coupe. And, I always had the feeling that the 'Couper's held a certain tacit respect for my flying skills, especially since the Pacer was as much a curiosity as their airplanes. It was almost as old, and at the time, in the mid eighties, there weren't more than a couple hundred production-model tail draggers of any age still flying.

And it's true that flying tail-wheeled airplanes requires special training. All airplanes may fly pretty much the same–it's just during the landing that things are different and get touchy. And most of the required training teaches the pilot to keep the airplane from ground-looping, during the rollout portion of the landing.

This maneuver added risk to every landing because the 'center of lift' is located behind the main landing gear, giving the airplane a tendency to 'weathervane' during the rollout sequence. To counter that tendency, the pilot must be constantly alert during the landing and taxi operations. As long as the engine is running, the potential for a ground loop is always there.

Before I bought the Pacer, I worried about how dangerous it might be to own a tail wheeled airplane. But after talking with a high time pilot who'd flown tail draggers for most of his life, I realized that it didn't have to be a problem, if I got the proper training. He told me that most pilots who fly tail draggers and are either afraid of them or don't land them very well have never taken the time to properly learn how to land tail wheeled airplanes. There's a certain technique to landing them well, and if you never learn it, every landing will be 'iffy.' And after a while you'll be so afraid of the airplane that you won't want to fly it. Those pilots who always land on the main gear only and not three point, are the ones that never learned.

'If ya get a proper checkout and really learn how to land 'em, ya'll see that the airplane's really docile,' he told me. 'But,' he cautioned, 'Ya really have to get to the point in your training, no matter how long it takes, where ya know every second what the airplane is doing or is about to do. Anticipation is a pilot's preoccupation in a tail dragger.'

●

The Sunday morning of the Oceano fly-in was ushered in with bright sun and a clear sky. Les and I arrived at the airport early, and by 6:45 a.m., I'd already preflighted the Pacer, having topped the tanks the afternoon before. So with not much left to do but wait for Les, I started the Pacer and taxied across the field to where the Cadet was parked.

Les took his time doing a 'walk-a-round,' and just when I thought he was finished, he started the engine and taxied down to the wash rack to hose the dust off. It was another ten minutes before he was ready to go. I guess his ego wouldn't allow him to show up at a club fly-in in a dusty airplane. His airplane was a novelty to him, his pride and joy.

It was 7:20 a.m. by the time we got to the approach end of Runway 25, waiting for the Tower to clear us for takeoff and a 'box climb' to two thousand five hundred feet, the altitude for the VFR Corridor that tunneled through LAX's TCA airspace.

As we sat there waiting, I looked down at the departure end of the airport, and almost as if on cue, the fog was already beginning to form. If we didn't take off in the next several minutes, the fog would certainly ground us for the next three hours.

Les looked over at me through the Cadet's canopy and raised his hands as a gesture of frustration, shaking his head. I nodded in return, and looked down the runway towards the Tower. Finally, after what seemed a long time, we were cleared and I followed Les in trail, taking off and climbing a half mile behind him through the box pattern to Corridor altitude.

By the time we were to the Corridor, I looked down to see how the conditions were on the runway below, but all I saw was a blanket of fog that already obscured the field. From 'severe clear' to IMC in less than four minutes.

I called Les over the radio on the 'plane to plane' frequency, as he flew into the Corridor, and I told him to look down at Hawthorne's runway. I said it was a good thing that we got out of there when we did, or we'd still be sitting there.

'Yep,' was all he grunted in reply.

•

The Oceano County airfield is one hundred and seventy miles from Hawthorne, and takes about an hour and forty-five minutes at our one hundred and three mile per hour cruise speeds, in a no-wind condition. With a headwind, it could take maybe another twenty minutes. And it was my experience from previous flights, that by mid-morning a quartering headwind usually develops north of Santa Barbara along Victor 25. So when I travel in that part of the state, I always expect wind, and if it doesn't materialize, so much the better.

And from our previous fly-in experience with the 'Coupes, it's important for us to arrive early; they have a curious habit of arriving early and departing early — right after lunch. So if you arrive late, and just in time for lunch, then its certain you won't get to visit much before everyone leaves. After lunch you'll find yourself all alone, and talking to the trees.

To counter that elusive departure tactic, Les and I always arrive early and then take our time departing when most everyone else was gone

By 8 a.m. we were just flying over the Camarillo VOR station, having turned northwest from Santa Monica, cutting across the Santa Monica Mountains to the 101 Freeway. And, once over the San Fernando Valley, the weather was clear and sunny with a visibility of over a hundred miles. Forty seven nautical miles ahead, we could see the peak of the Santa Ynez Mountains where the San Marcus Pass cuts through the summit, just west of the San Marcus VOR station.

As we flew along, I realized that the wind never materialized, making the flight extremely smooth. And, for the most part, the radio frequency was quiet, only an occasional airliner could be heard reporting onto the 'profile' to Los Angeles

Approach. And once they reported over the radio, I looked up, high in the sky, and could see a silver body reflecting the sun's glint, as it proceeded to LAX.

Our cruising altitude was four thousand five hundred, westbound, an altitude that allowed us to clear the highest peak of the San Marcus Pass that had an elevation of four thousand two hundred and twenty nine feet. The clearance was close, but on a quiet morning with no wind gusting, two hundred and eighty one feet was enough.

We flew along in silence since Les doesn't like to talk much over the radio. He never explained why, and I never asked, but our flights were all pretty much the same; either total silence, or me asking the questions and verbalizing the situations, and then Les grunting his replies. But as we approached the San Marcus Pass, Les caught me off guard by calling me over the radio.

'Bill, look at your one o'clock, and three hundred yards,' he said, excitedly.

'Got 'em,' I replied, looking twenty degrees to my right. There appeared an Ercoupe at our altitude joining up on Les' right side. It was painted silver with red trim and looked like it was show quality, as it flew along, its finish glistening in the sun.

'Ya think he's goin' to Oceano with us, to the fly-in?'

'Don't know!' We'll see how long he stays with us.'

Approaching the Pass, I went back to scanning the sky for other airplanes, but soon the mountain peak climbed closer to us for a brief moment, and then fell quickly away, as we crossed the summit and continued along the northern slopes of the range, angling towards Lake Cachuma. I was watching the southwesterly sky towards Santa Barbara, ten miles off my left wing as we traveled beyond the Pass. As the mountains began to obscure that area, I returned my gaze forward, looking for the Ercoupe that suddenly had disappeared.

'Where's the 'Coupe?' I asked over the radio, surprised that it wasn't there.

'It's below us, and falling behind. I think he's going into Santa Ynez.'

'Okay. I see him now,' I returned, looking over my right shoulder. The 'Coupe by now was five hundred feet lower and at my four o'clock and a half mile position. It was heading directly towards the center of Lake Cachuma, and would probably follow the Lake to the Santa Ynez airport, as it bled off altitude.

•

The airfield at Santa Ynez is at an elevation of six hundred and seventy one feet, with its single runway just under three thousand feet long. At altitude, to clear the Pass, an airplane has to begin an immediate descent so it can arrive over the airfield fairly close to pattern altitude; otherwise, a pilot can find himself several thousand feet too high. And since flying over the lake is fun, most pilots like to descend over the fishermen and sail boats on the way to the airfield.

The Santa Ynez airfield passed below us, as did the town of Solvang several moments later. Remnants of the morning fog still hung over the valley in thinning whisps, mostly above the meandering river.

The valley extends to the west from Lake Cachuma to Lompoc, a distance of some thirty miles, and no matter what time of year, the view is always magnificent. Even though the valley is only six miles wide, what makes it so picturesque are the many farms nestled in the foothills on either side of highway 254 that bisects the parallel mountains on either side, running westward, the length of the valley from the Lake. The

Stanta Ynez Valley is one of my favorite places to visit either by airplane or car.

•

Once beyond Solvang, Les turned to the northwest, on a heading of about 307. It turned out that he wasn't pointing at anything in particular, but was homing on a white patch on the horizon, forty five miles away, that was Pismo Beach. On a clear day or early in the morning, before the wind stirs up all the dust and haze, the distant white patch of sandy beach is easy to see from cruise altitudes.

Like bush pilots flying the North, once you've flown to a place and know what the landmarks are, you really don't need a map to find it again. And the landmarks become the sign posts by which pilots measures their progress.

The area between Solvang and Santa Maria was not visually familiar to me; most of my flying north of Solvang was on Victor 25, an airway some twenty miles further east from our present course, where the terrain is mostly low rolling hills with an entirely different look. But as we neared the airport at Santa Maria, the coastline began emerging as we flew on toward the distant white patch. And slowly, as the distance to Oceano shrank, the terrain became very reminiscent of the Moss Landing area of Monterey Bay. It also has an Edison plant and tall chimneys and various inlets and estuaries along the coast.

Another quarter of an hour and we were just overflying the airfield at Oceano. I stayed behind Les and followed his maneuvers. He came in low over the town and then crossed the airfield at the mid-field point, turning downwind over the acres of artichoke and lettuce fields. Finally, turning base and then final, as I followed at a distance.

As I overflew the airfield, I noticed that although the runway wasn't very long, at least it didn't seem to have the potholes that other small strips had, from years of inattention, like the runway at Agua Dulce, near Palmdale.

The elevation of the field was only fourteen feet above sea level, and the ambient temperature was always a cool Standard Day fifty nine degrees F. It seemed like a great place to compare an airplane's indicated airspeed against its book figures, as no interpolation was necessary at fifty nine degrees F. And for a moment, I considered doing just that, but then dismissed the impulse, as the performance of a bush plane like the Pacer was never a concern to me. The Pacer was a 'special purpose' vehicle, and it's value is in its utility, not in its performance. It's like owning a jeep and worrying about how fast it goes — not really it's purpose in life.

After landing, I followed Les to the transient parking area where one of the Club members was standing, ready to direct arriving airplanes into parking spots. Even at this early hour, there were already two Ercoupes there, with four members grouped in front of one of them, discussing things.

As soon as Les finished securing his airplane, the others flocked to his Cadet. It was the only one still flying in Southern California, and 'Coupe pilots were always curious as to how it was different from the standard 'Coupe.

•

The airfield itself was uncontrolled, but a Unicom frequency was available for airport advisories. The interesting thing was that the airfield was located only one block from the outskirts of the township and only fifty yards from the sand dunes of Pismo Beach. The beach was directly ahead of the departure end of Runway 29, so that all flights taking off that way crossed the sand and climbed out over the ocean. And, although the field itself was small, it seemed to be big enough to support two FBOs. Beyond the large hangar buildings that housed the FBOs, a dozen permanent airplanes were parked on the ramp, while a grouping of portable 'T hangars' lined the field on its north side.

After we landed and tied down our airplanes what begun as a group of four pilots grew an hour and a half later to a crowd of twenty five — with twenty two airplanes parked side by side along the ramp. The sight of that many airplanes was impressive. And for a small airfield that never saw more than a couple of dozen airplanes land through the week, hosting a fly-in there seemed to be appropriate and some how comfortable.

I sat down at a picnic table next to the only FBO that was open and whose sign read: 'CONDOR.' The other FBO next to Condor was closed on weekends, although by the looks of it, it could've been closed permanently. It had no sign.

As 'Coupe after 'Coupe buzzed the field and then landed, it was really interesting to see the same models each painted differently with a variety of modifications and configurations of instruments. When the last airplane had landed and everyone had arrived, there were twenty two airplanes on the transient ramp: twenty one Ercoupes and one Pacer, mine.

Condor's fuel truck provided line service and was operating continuously during the two hours that the airplanes were landing for the fly-in. It went from one airplane to the next, right down the line, as each airplane had its tanks topped for the trip home.

I got to talking to one of the locals who wandered by and discovered that just a year before, a middle aged Swede bought the airfield and started the Condor FBO. He was blonde and blue eyed, a rather Nazi Nordic type, complete with tan — not quite the pasty complexion of an 'SS' trooper. Actually, he looked pretty much like a ski instructor from Vail. And although he'd been in the States for only a few years, he'd already lost his accent, and now talked like a Redondo Beach surfer.

And being enterprising, he'd built a second floor loft in FBO's hangar and made it his office. And sometime that morning I happened to pass the time of day with him and he told me that he purchased the airfield and then spent all of his money trying to keep up the FBO. He wanted to make it a paying proposition. And though Oceano was just a small airfield, and somewhat out of the way, his marketing strategies were ingenious and hopefully soon would payoff,

His most imaginative strategy was to hire two very busty young women CFIIs for his staff: a blonde and a brunette. And, no matter how long I live, the one memory of Oceano that I will cherish forever has to be about the brunette who drove the fuel truck that day, refueling all of the airplanes.

The brunette who provided line service for Oceano that week, by driving the fuel truck and pumping gas, was a flight instructor teaching student pilots their primary flight lessons. While she was on the ladder refueling my Pacer, I got a chance to talk to her, and she told me that she was working on her PhD degree at night from USC, in Los Angeles. I could tell that she was very clever in her marketing techniques and that obviously she knew that her 'fuel-buying' customers were male pilots. And, it was also apparent that she knew exactly how to ensure that they bought as much fuel as their airplanes could hold.

Since most of the pilots were male, she realized like Gypsy Rose Lee, that she needed a 'gimmick' to 'hook' her customers. And her gimmick was to wear a man's long sleeve white dress shirt, with no bra on underneath, and with the first three buttons unbuttoned, so when she stood behind the trailing edge of a low winged airplane, and leaned over while holding the fueling hose, at least one of her two breasts were guaranteed to be exposed. That ploy also worked equally well with high winged airplanes, except that she'd be on the top of a ladder bending over. She also made certain that she was facing the front of the airplane, where the pilot would be standing and couldn't help but watch, transfixed.

Since she was so obvious about it and made no attempt to conceal the spectacle, I decided to stand at the base of her rickety ladder and steady it for her, while she refueled the Pacer. From that vantage point it was obvious that I would miss the view and that I wasn't interested in ogling her front. Besides, she was already topping both tanks of the Pacer, what more could she sell me in the way of fuel?

Since I wasn't ogling her and she didn't have to concentrate on exposing herself, we ended up having a rather nice conversation about her job at the FBO and the business prospects for Oceano.

She wasn't particularly pretty, but then no one who bought fuel from her that day could possibly remember, since that was her one feature they weren't paying any attention to. She was quite a sight!

I did notice that Les rather enjoyed the show as she pumped fuel into his Cadet, holding the hose just so. The look on his face told me that he was quietly wondering which one of her breasts would pop out first. I think the funnier sight was watching Les watch her. It was one of those rare moments in life that was over too soon.

•

The other woman, the blond, turned out to be a talented baker, and displayed her pastries in one corner of the FBO.

Several tables had been set up displaying trays of homemade donuts, pies, cakes, and cookies. Each of the cookies were at least six inches in diameter, and a sign that listed the prices reminded everyone that although convenient, these pastries weren't cheap, which kept the sales slow for the first several hours, but as lunch time approached, the pilot's wallets gave into their hunger.

On a table next to the pastries was a coffee service. I tried a cup and really liked it. The coffee was strong and black and over the course of several hours I drank three cups, at a dollar a cup. Even without the pastry, the coffee was great.

The blond wasn't as personable as the brunette, and didn't bother trying to compete with her gimmick. Actually, her gimmick was that she was a fabulous baker. And since her coffee was the only coffee on the field, the coffee sales were brisk.

The Swede's hand in this ancillary venture was evident and his staffer's had done well in making money anyway they could — whether aviation related or not.

•

After most of the pilots and their passengers had looked over all of the airplanes and were finally talked out, they slowly moved towards the Condor hangar. And after a while, when all of the bakery goods were gone, except for a few cookies and one piece of apple pie, the group began to meander towards the town, two blocks away.

At the restaurant most of the pilots acted as if they hadn't eaten in days, although they'd just cleaned out the baked goods at Condor. Suddenly, it dawned on me why most middle-aged pilots aren't skinny.

That morning on the flight ramp, Les spent his time visiting one group and then the next, while I stayed in the background and just 'people watched,' until the word was passed that it was time to head over to the town's only Mexican restaurant for lunch. Slowly a group migration began in the direction of Grover City, the town adjacent to the airfield. From the airfield, the town began just a hundred yards away — about two short city blocks. And since the city had only five or six streets each way from the town's center, it didn't take long to arrive at the restaurant on foot. Inside, the waitresses pushed six tables together to make one long one, where we could all sit together. The people as a group were certainly friendly enough, and there were always several conversations going at once. Surprisingly, the food was better than just good. Wayne Olson, the tacit leader of EOC's Region 7, AKA 'Mr. Ercoupe' and his cronies, sat at the far end of the table and were engrossed in a spirited political discussion about the Regional EOC Club elections coming up soon. Les and I sat at the opposite end and got involved with talk of a much more mundane nature.

The entire lunch lasted an hour from start to finish. The 'Coupers were eclectic to say the least. What was curious about every EOC fly-in was that as soon as Wayne Olson stood up, that was the signal that lunch was over. Then, it seemed to be a race to see which member's airplane could become airborne first.

When lunch ended and the main body returned to the airfield, several 'Coupes had already started their engines and were ready to leave. Les and I meandered back to Condor, talking to several people enroute. The urgency for Ercoupe Owners Club members to depart directly after lunch was unique to them. Other type club fly-ins that we attended gathered afterwards in small groups, as those members seemed sorry that the afternoon was over and were almost reluctant to leave. I think the 'croupers' were pretty much all talked out by lunchtime. It was almost as if they felt awkward to stay and make small talk. So by the end of lunch their socializing was done.

I thought about this curious custom and finally decided that the only thing any of them had in common was their Ercoupes. And after several hours of talking about that subject, it was as if they didn't have anything else to say. They seemed to be a 'people' who were shy or untrained in social skills and felt awkward when conversations strayed from airplanes.

By the time Les and I returned to the airfield, the taxiway was jammed with airplanes, all waiting their turn to escape. The 'Coupes looked like a squadron of Navy fighters on the flight deck of a carrier ready to launch.

I looked over at Les, and motioned with my head towards the jammed taxiway. He looked back, shaking his head, laughing.

•

The flight back to Hawthorne was uneventful, and as we cruised along southeast bound over Solvang, Les unexpectedly turned easterly, paralleling the Santa Ynez Mountains on the north side, instead of crossing the Pass and following the coastline back from Ventura. I guess he was always leery of flying the coast, since the likelihood of meeting oncoming traffic was too great. So we continued on an easterly course through the San Fernando Valley to Topanga Canyon, then flying over the low coastal hills of the canyon, breaking out over the beach two and half miles west of Santa Monica Airport and the VFR Corridor that transited LAX.

•

I flew in trail behind Les as we went through the Corridor

over LAX. Les was always in a hurry, flying his airplane full throttle. I guess he was impatient to go places and then get home. I'm just the opposite, feeling alive while I'm flyIng and in no great rush to arrive anywhere. Just being up in the sky and flying was where the fun was for me. I never really had to go anywhere, just up. Flying for me was visiting my favorite place, the sky. I always looked forward to it, even though it was just for a visit.

One by one, we landed back at Hawthorne. Les taxied to his parking spot and I to mine. We said our goodbye over the radio. A brief 'So long, see ya,' was usually all it took.

Hesperia Air Lodge

For some time prior to Les' decision to buy a Mooney Cadet, he and I spent considerable time in a high desert town called Hesperia, located thirty miles northwest of San Bernardino, just south of Apple Valley, where Roy Rogers lives. In fact, we made four trips there in my Pacer, within a six month period, so that Les could visit and commune with one of the last Guru's of the Ercoupe and its derivatives, Wayne Olson. Les would seek his counsel about various models, as he tried to figured out which airplane he should buy

Every time Les and I flew out there, I always looked forward to it. Wayne's house was located right on the taxiway of the Hesperia Air Lodge. And it was always a good feeling to know that some pilot actually was living the 'good life,' just in case I would never be able to. At least, at Wayne's, for a moment, I could experience what living on an airport, with my airplane parked under its own 'car port' in the back yard, was like.

Wayne Olson would never tell me exactly how old he was.

He would only admit to being over seventy. When I first met him, he was probably in his early seventies, but seemed much younger, a dynamo of energy. He was tall, well over six feet, with thick gray hair, a large frame and a deep voice–a really great voice; like an FM Announcer. In his youth, he was probably considered handsome. And now, on a good day, he reminded me of Loren Greene — Ben Cartwright on the TV western, Bonanza.

As the unofficial guru of Ercoupes, Wayne's claim to fame was, that in his time, he'd bought and owned seventeen airplanes: all of them Ercoupes or derivatives of Ercoupe, like the Mooney Cadet. There was no one around for two states either side of California who knew more about the Ercoupe than Wayne. Having owned so many in one lifetime is no little feat for a pilot. It would usually take the average flyer, if he could even afford it, several lifetimes to go through so many. But, because of Wayne's longevity, he'd been flying them since the late thirties. By the time I first met him he'd already been flying for fifty years, which made him a really special person.

He knew every nut and bolt and hose fitting by their part numbers. But as he got older, he told me that it got harder for him to work on them, because of their compact size. The 'Coupes were always small, but he kept growing. And as anyone knows who's ever looked behind the instrument panel of a 'Coupe, there isn't much room for any additional stuff, like modern electronics. But, all contemporary owners seem to moderize their planes by adding additional electronics to their instrument panels.

In 1936, when the airplane was first designed about the only flight instrumentation required was an airspeed indicator, an altimeter and a compass. The other two or three instruments on the panel had to do with the engine: like the oil temperature, pressure gauges, and the tachometer. And, as Wayne aged, his large hands had trouble working in such cramped quarters; and, instead of it being a labor of love, it was more like an aggravation.

Not only was there little room behind the instrument panel, but forward of the firewall, in the engine compartment, there wasn't an inch to spare. And modern flyers are always trying to 'improve' their airplanes.

Anyway, prior to Les' decision to buy a Mooney Cadet, we frequently flew into Hesperia just so Les could discuss with Wayne the various attributes of the Cadet. Of course, this was long after my 'love affair' with the Ercoupe had cooled, since I was too big for it. Les, on the other hand, was shorter and lighter and had no problem with the interior size of its cabin.

Les was very anxious to buy his own airplane, and of course, the airplane he really wanted, but couldn't afford, was a Mooney 21, the same high performance model he'd spent his flying years renting. The price was five times more than that of a Mooney Cadet, but he vowed that someday he would get his '21.'

I remember the last time that I visited the field at Hesperia. It was at an EOC Fly-in in April, 1990 and I remember that I had to drive in my car to get there, because a low overcast had socked in the entire Los Angeles basin the night before. And although the event was a two day thing, I was only going to show up on Sunday; whereas, most of the fly-in attendees arrived on Friday, for the Saturday-Sunday affair. The weather on Friday happened to be VFR. Of course, once over the mountains and in the high desert, the weather is always VFR. The problem is getting there from somewhere else.

It was a regional fly-in and a lot of airplanes arrived from all over the southwest, as well as from all the states surrounding California; like Arizona, New Mexico, Oregon and Nevada. It was a huge success. There were thirty-one 'Coupes in all, as well as several other non-Ercoupes.

I drove there because I really didn't want to miss it and I wasn't about to get 'weathered out.' When I got there, I was surprised to see Les' Cadet and his friend Paul Hamilton's Messerschmidt 190 parked on the ramp. They of course must've arrived several days before with everyone else.

Wayne Olsen was there holding court besides being the unofficial Master of Ceremonies. The lunch was held in the factory building of one of the light manufacturing companies located next to the north taxiway, just across the runway from Wayne's house. The owner of the factory was also a Club member and volunteered his facility. On the ramp in front of the building were airplanes parked wing tip to wing tip for almost a half mile. It was quite an impressive sight.

Most of the flyers were engaged in discussion groups of five or ten, talking about their Ercoupes, while inside the building's lunchroom a half dozen women set up a great buffet, with hamburgers, hotdogs, potato salad, and baked beans. The eating was fine.

Les and Paul and I sat at a table inside, and I listened to them tell me about all of the contests that had transpired the day before, while I ate my lunch. There were spot landing and flour bombing contests. The only contests they didn't hold was a poker run.

Sitting near us at the table was a middle-aged man who had been driving by the airfield when he spotted the Ercoupes parked side by side, and he stopped to see what was going on. He said that the Ercoupe was his favorite airplane and that he'd always wanted to learn to fly, but could never afford it. It turned out that he was a local high school teacher in Hesperia, really interesting to talk to, and quite knowledgeable about the 'Coupe. He said as far as he was concerned, the Ercoupe was the perfect airplane.

After lunch, prizes were awarded to the winners of the contests, and after the awards a raffle was held for about a dozen different gifts. The fly-in itself was really a well supported event.

•

Now, although Les and I still go to fly-ins, now and then, we've never had an occasion to return to the Air Lodge at Hesperia. After several years of owning his Cadet, Les moved up to his dream airplane, a Mooney 21, and I moved up to a Beechraft Musketeer. Once a pilot moves up to higher performing airplane, the fun of hanging around with 'Coupers just wasn't there anymore.

And just a couple of years later, Wayne Olsen lost his medical certificate because of poor health and had to stop flying. His doctor told him that he could easily get it back, but first he'd have to give up smoking and then lose some weight. But, after a lifetime of flying and messing around with Ercoupes, maybe Wayne just decided to 'call it a day.'

Santa Ynez

Fly-ins are fun if you're an attendee, but one time I had the dubious distinction of hosting a fly-in for the members of a national 'type club' that I belonged to called: The Short Wing Piper Club. Flying into a gathering of friendly pilots, all of whom own the same kind of airplane and enjoy the amenities, is easy and fun for everyone except the person stuck hosting it.

Once you've seen fly-in gatherings from both sides, it's always much better being a guest than a host. The following is my account of a fly-in that I was stuck hosting. It was the one and only time, but I'll never forget it.

I'd been a member of the Short Wing Piper Club (SWPC) ever since I bought the Pacer. The term 'short wing' has to do with the fact that the entire series of airplanes that Piper Aircraft produced since the close of World War II, had much shorter wingspans than Piper's famous and ubiquitous J-3 Cub And the Pacer was the second last model to be produced by the company with the configuration of short wings; the Piper TriPacer was the last in the series.

The SWPC was a national organization of some 1700 members at the time (now it's over twenty seven hundred) throughout fifteen regional chapters of the contiguous States, plus Alaska. The newsletter of the Club, the *Short Wing Piper News*, was the 'house organ' publication that averages over one hundred pages per issue and was pocketbook sized and stitched bound like a paperback novel. At the time, It was probably the best newsletter of any airplane type club in America.

And if someone owned a classic Piper airplane, meaning one built before 1960, the only way to keep up with maintenance and places to find production parts was to join a club. The Club's newsletter alone was worth the price of the Club's dues, since it always had articles that addressed difficult problems as well as advertising hard to find parts.

One of the Club's national directors cornered me while I was attending the EAA's Copper State Fly-in in October, one year, at Casa Grande, Arizona. He said that their records indicated that in California there were at least twenty Short Wing Piper owners that belonged to the national club. And it was his contention that if regional chapters of the national club weren't formed where locals could get together and support each other, then after a while, they'd drop out of the national club. He felt that since California didn't have a regional chapter, twenty some people would drift away if we didn't form one soon.

To remedy the situation, he wanted me to host a California fly-in to determine how many people would actually show up and be interested in forming a regional chapter. He was quite a salesman and told me that it wouldn't cost me much, just the price of mailing out postcards to the twenty or so members. I don't know why, but I believed him.

The post cards were to announce a place, time, and date for the fly-in. Actually, in the end, it cost me a lot more than just postcards. I also knew it would be a lot of work as I'd been involved in social gatherings like this before. But I knew he'd think I was deserting the national club if I didn't say yes, so I did.

From previous experience with hosting meetings, I was certain that once I mailed out the postcards requesting an RSVP, nobody would respond. That seemed to be human nature. So, I knew that in the end, I'd really have no way of knowing how many people intended to come, if any, expecting to be fed and entertained.

But, being a good 'soldier' and accepting the assignment (knowing full well that there was every possibility that no one might show up) I continued my efforts at producing the event.

The national club gave me the addresses of Californians that owned short winged Pipers, and as it turned out, most of them lived in Southern California. So by the time I was done with the mailings I'd sent out thirty postcards. And, as I thought about it, for a group as large as thirty-five, I needed to hold it at an airport with decent facilities in a picturesque setting, so that pilots might be attracted to attend just because of the area.

The only airfield that qualified was halfway between San Diego and Sacramento. Santa Ynez my favorite place, a small country airfield with a single paved runway, perched on a plateau in the center of a sleepy valley, just north of Santa

Barbara on the other side of the mountains. It seemed the perfect place. Even if the pilots didn't want to form a chapter, they might want to fly-in just for a pleasant afternoon. If the weather enroute to Santa Ynez was halfway VFR the people would come, since, the weather at Santa Ynez was always 'severe clear.'

In preparing for the fly-in, one of the people on the list was based at Santa Ynez and was also the current President of the local EAA Chapter there. His name was George McNeely, and as soon as he received my postcard he gave me a call, offering the facilities of the EAA hangar for a barbecue and meeting. When I got the call, I thought I'd died and gone to Heaven. News that good right out of the blue was like manna from heaven, since prior to the call I had no idea of how or where I was going to feed the people if they showed up, or where I was going to hold a meeting.

George told me not to worry: he'd hosted dozens of fly-ins at Santa Ynez, since he'd been based there. He told me it wouldn't really be much work for either of us. And, although I didn't believe him, in the end he was right. He said that besides getting the barbecue ready, all we really had to do was buy the food at a nearby grocery on the morning of the fly-in. It turned out that he had all of the tables and chairs we would need, as the EAA meetings required the Chapter to feed and seat about forty people at their monthly meetings.

I began to relax a little after George's call, especially after he told me that he'd give me a hand. And, as the days and weeks passed and the event neared, just as I suspected, nobody responded to my RSVP. And if that wasn't enough of a problem, just the Saturday before the fly-in, my Pacer developed a fuel leak in its left wing root. It later turned out to be just a thirty four year old rubber hose connection that had decided to begin seeping fuel onto the headliner.

The hose connected the aluminum fuel tank in the wing to the aluminum fuel lines that ran from the wing root to the fuel selector valve behind the kick panel near the Pilot's left knee. It was an easy fix, but getting at it meant ripping out the entire original headliner. And one thing led to another, and by the time my mechanic and I were done, we'd taken out the seats, side panels, kick panels and carpets. We decided that as long as we were going to fix one hose, we might as well replumb the entire fuel system, especially since it was over thirty years old. So, needless to say, my Pacer wasn't going anywhere soon.

My problem was that I was hosting the fly-in and I really felt compelled to fly into it. Wouldn't it look silly for me, a pilot, to drive the three hours to Santa Ynez from Los Angeles to my very own fly-in, when it was just an hour by air?

So, on the day of the fly-in, I rented a Cessna 172 at the Torrance Airport, and by 8 a.m. that morning, I was winging my way to Santa Ynez. The Day promised to be particularly gorgeous, with a clear blue sky and bright sun. It was the first weekend in more than a month, when the weather seemed to hint at the coming of summer. For the past six weekends, the weather had been the usual spring time 'late night and early morning fog,' the kind that lingers over the coast until mid-afternoon. But today, the weather was crystal clear, and I took it as a good omen.

I had the throttle of the Cessna fire-walled as I flew along, but it didn't seem to help the speed. The Skyhawk was a hundred and twenty knot airplane, and it was going as fast as it could, 'rain or shine, mud or flood.' And since the distance to Santa Ynez from Los Angeles was one hundred and eighteen miles, it took a solid hour to get there. So, by 9:30 a.m. I was parking the airplane on the transient ramp of the airfield.

Next to the office building was the car rental office, so I walked over to it and took delivery of the Lincoln Continental that I'd reserved for ground transportation. I thought with a big Lincoln I could really haul people in style from the transient parking ramp to the EAA hangar.

I immediately felt better once I got the car and drove the length of the field to where the EAA hangar was located. I met George and the next two hours were rather hectic. George and I made several trips to the local grocery, three miles down the road from the airport. We set up tables and chairs for forty, and finally ignited the barbecue to begin heating the charcoal for cooking the hot dogs and hamburgers.

By 11:30 a.m. everything was in readiness. George and I sat down to rest and wait. We were hopeful that somebody would eventually show up. A half hour went by, and then another. No one showed. George told me not to worry, that they'd come.

But, as the time passed, I wasn't so sure. I paced the hangar and stared at the empty sky. Then suddenly, a little after noon, one by one, Short Winger's began to appear. And then for another hour there was a constant stream of classic Pipers landing and taxiing up the the EAA hangar, parking, and lining each airplane up with the one beside it.

Except for four airplanes, all of them were tail draggers. There were Cruisers, Clippers, Vagbonds, Pacers, TriPacers, my rented Cessna and a Beechcraft Bonanza.

Seventeen in all. By the time all of them had arrived, I was frankly amazed, especially since no one had responded to my invitation.

People came from all over: from as far south as Bonita, California–which is about as far south as you can get, without living in Mexico; and as far north as Happy Camp, California–about as north as you can get, without voting in Oregon. But mostly, people came from Fresno and parts south. And there were at least a half dozen from the Southern California area: like Whiteman Airpark El Monte, Long Beach, and Compton.

By the time we had the barbecue and fed everybody, there were thirty-seven people for lunch. And once lunch was finished, my part of the program was finished. I had recruited a really dynamic pilot from the San Diego area to run the business part of the get together, so, I took a seat in the back row of the hangar, relaxed and watched and listened to the proceedings.

By the time the afternoon was over, the group had formed the Southern California Chapter of the Short Wing Piper Club, and even elected officers for the year. I thought that creating a regional chapter before my eyes was simply amazing. And when the business meeting was over and I hadn't been elected to any office, I was especially glad.

By fly-in standards, the gathering was a success. And I was smart enough to realized that the only reason most of the people came, was not because they were eager to form a regional chapter, although they did, it was because of the great weather. Good weather is always an attraction to flyers–you can't keep pilots on the ground when the sky is clear and sunny.

Chapter Four
It's A Long Way To Tipperary

Bringing Back Your First Airplane, Alive

It was towards the end of March when my friend Les called me at work one afternoon to tell me that he was about to buy an airplane. I'd already owned my Pacer for over a year, and I knew that secretly he was envious of my being first to own an airplane. He'd been renting airplanes for more than twenty years, before I even learned to fly, and I suppose he felt it wasn't fair. If I could finance a vintage airplane, why couldn't he? So I think he called in some IOU's to finance an airplane of his own. He never mentioned his displeasure, although he'd been flying most of his life, and I'd only had my license for twelve years.

I always regarded Les as the consummate Sunday Pilot, as he'd been flying since his Boy Scout days: first gliders and then powered airplanes. And he was what many would call 'a natural.' I remember one time during the late seventies, he stopped flying for three years, and then suddenly got motivated to get back in the air. To get re-qualified, all he did was take an FAA physical and then went down to an FBO to take a Biennial Flight Review. Not having touched an airplane for such a long time, he was worried that he'd lost 'the rhythm' and would require a few hours of air work just to get current. But, the CFI who flew with him signed him off after just a half hour. By contrast I laid off from flying for a year and a half while going to grad school, and it took two flights and almost four hours of airwork before I felt comfortable again in an airplane. Of course, I'm not a natural like Les.

Anyway, Les phoned to tell me very excitedly that he'd finally found the airplane of his dreams–the one he'd been looking to buy after months of searching, and after a half year of communing with Wayne Olsen about various models. Les had decided to buy a Mooney Cadet, a derivative of the Ercoupe design that Mooney Aircraft developed after buying the manufacturing rights from Ercoupe in the sixties.

For a dozen weeks Les searched various Cadets, finally locating one that he thought had his name on it. He hadn't seen it yet, since it was located in Fort Meyers, Florida. But, I guess he spent hours on the phone talking about it with its current owner and was convinced that it was the right one to buy. Of course, the main reason it was most attractive to him was its price. Since the price was right, he could now buy his airplane and pay cash.

He told me he was so serious about it that he'd already cut a cashier's check for the specified amount. And then, he said that he was going to take two weeks vacation and fly down to Fort Meyers and check it out. If it's condition was as advertised, then he was going to buy it on the spot and fly it back to Los Angeles.

'What if ya get down there and the airplane's a dog?'

'Well, I'll just come back with my check,' he replied, as if the matter was settled.

'Well, when are ya going?' I asked, thinking about the fun of riding back with him in his newly acquired Cadet. It could be a real adventure.

'I think I'll leave in about a week or so. I'm just going to take the vacation and go down there and do it,' he rambled, listening to his own words, hoping that hearing them out loud would make them come true.

The whole thing sounded awfully tentative, but I could tell by his tone that he was serious and nothing would stop him from getting his 'little Mooney.' And, since he was going to go anyway, I didn't want to miss out on a flying adventure if he actually bought the Cadet. I figured he could call me from Florida when he'd start back to Los Angeles and we'd meet halfway. I'd meet him and ride back with him in his own airplane. I thought it would really be a lark.

After a lengthy chat, he agreed that I could ride back with him from Austin, Texas, as that was about the halfway point.

'I hope I buy the airplane,' he said over the phone, 'cause I've already planned the entire flight from Florida. 'There's an airport ten miles north of Austin's Municipal field, called 'Executive Airport.' It's an uncontrolled corporate field. I'll be stopping there for fuel. So if I do buy the airplane, I'll give call ya in plenty of time to get to Austin.'

•

And so, on a Saturday morning Les jumped on United Airlines and departed for Florida. He arrived to discover that the seller of the Cadet, Bob Duffy, was a man of means and had a fairly large estate twenty miles east of Fort Meyers. Les spent the next several days there, seeing the estate and getting checked out in the airplane.

He told me later that Duffy owned thirty acres of sprawling grounds of lush shrubbery and lawns surrounding his large Mediterranean villa, as well as a three acre pond located on a large expanse of undeveloped land, some distance from the house. And beyond the pond, Duffy had built his own private grass airstrip.

The afternoon that Les arrived was spent touring the property, finally ending up at the three acre pond. The pond was a large irregular shaped oval, with a dozen small palm trees, knee high grass and low bushes growing down to the water's edge. It looked like an oasis. It was there that Duffy showed off his prized pet — an alligator named Henry. The alligator was ten feet long and Duffy had raised him from a baby. Henry was sunning himself at the water's edge near a thick bed of saw grass.

'What do ya feed him?' Les asked.

'Nothin' now. Thair's so much wildlife aroun' tha pond, I stopped feedin' him. There's plenty for him to catch on his own, raght hair.'

When they'd finished the tour and were back inside the villa having a drink, Les casually asked the owner about the snake situation, curious that during their walk they hadn't seen a single one.

'Oh, thair all ovah, but mostly roun' the pond, this time ah day, 'Duffy said, slowly sipping his drink.

'Ya mean the pond— where we were?'

'Yassah. It's mighty snaky down thair. Lots of 'mocks,' he nodded.

'Mocks? What's a Mock?'

'Yawl hair 'bout 'Cottin Mouth Wattah Moccasins?'

The second Les heard those words, he told me later, his heart started racing and his knees began trembling. It was an uncontrollable reflex. He couldn't stop his knees from shaking, because the one thing in the world that scared him to death were snakes. He remembered the pond and its overgrown grass and shrubbery growing down to the water's edge.. Both he and Bob had just been standing in knee high saw grass for more than a quarter of an hour, as they watched Henry. Their legs were probably within inches of large ugly Water Moccasins. Just the thought of standing knee deep in poisonous snakes gave Lees a fright.

His fear turned into panic and was so strong that he felt like getting up and running from the house. He couldn't believe that they'd escaped being bitten. The Cotton Mouth was among the most aggressive and territorial of any snake alive. They've been known to chase men through their territory, not stopping until the chase ended in either their death or exhaustion.

Bob, being a southern gentleman and not wanting his guest to feel alarmed, took Les' glass and refilled it with whiskey and several ice cubes, and handed it back to him, nodding for him to drink it. 'It'll calm yawl. I'm sorrah if the mocks' made yawl uncomfortable.'

Les could feel his heart in his throat and it was many minutes and another whiskey before he calmed down.

•

The private airstrip that Duffy had built for his Cadet was twenty five hundred feet long and fifty feet wide. Its edges were outlined like a football field with white boundary lines of lime. A broken white centerline was drawn down the length of the runway. And since the airfield was twenty miles northeast of Fort Meyers Airport, there wasn't any concern about a conflict with other traffic; It was the perfect private airport.

The Cadet turned out to be in excellent shape mechanically;

cosmetically, it needed new paint, but that was all. The seller was sorry to lose it, but he really needed the money; he'd recently retired from his own company, which manufactured recreational vehicles. After a year of retirement, he decided to start up another business. But, he needed some front money before the bank would give him a loan, and the asking price of the Cadet was the exact amount to float his new venture.

Duffy told Les that he wouldn't be grounded for long, since he planned to buy another Cadet as soon as his new business was established. Duffy thought it would only be a year and he'd have another Cadet. It was his favorite airplane.

After a lengthy inspection of the Cadet and its logs, and satisfied that there was nothing that needed fixing, Les handed Duffy the cashier's check and took delivery of the airplane.

For most of the morning of the second day in Fort Meyers, Bob sat in the right seat, as Les practiced takoffs and landings, as well as the usual airwork that one does during a good checkout in an unfamiliar airplane; like stalls, slow flight, and emergency procedures. Since Les was such a natural, after two hours Duffy let Les fly solo for the rest of the morning.

By that afternoon, Les felt confident that he could fly and land the Cadet without doing any damage to it; after all, he had his CFI license and more than ten thousand hours of flying time under his belt. But, to be honest, the humidity of the Gulf Coast and characteristics of the grass runway were new factors that Les had no experience with, and had to adjust to; after all, the performance of the Cadet's 90 horse engine was really affected by the heat and humidity.

During takeoffs, the climbout seemed to be dreamlike affairs, as well as the takeoff roll. But, after a few hours of flying the Cadet, Les fell in love with it. It wasn't a Mooney 21, but it was his and he couldn't wait to get started the next morning on his journey back to California.

With the Cadet's cruising speed of one hundred and ten miles per hour, the two thousand three hundred miles from Fort Meyers to Los Angeles should take about twenty one flying hours, in a no-wind condition. But, flying from east to west in the northern hemisphere, meant that there would always be some headwind to slow an airplane down, due to Correolis Effect. It's just a matter of how much headwind. Les figured it would probably take about thirty flight hours to get back home–which could be completed in about four days of good flying weather.

On the morning of the third day of his two week vacation, Les planned to depart. The day before, while he was busy doing 'touch and goes,' while getting the feel of his new airplane, he noticed that the Florida weather was different than California's. The skies would cloud up for a while, then the clouds would dissipate; and sometimes a rain shower would come along for twenty minutes and then disappear, and then the skies would clear and the sun would come out for the rest of the day.

But Bob Duffy warned Les that as benign as the weather appeared, the spring thunderstorm pattern of the Gulf Coast was nothing to trifle with in a small airplane, or airplane's of any size, for that matter. He lectured Les on how unpredictable and stormy the weather could be.

But, Les thought to himself, how stormy could it really get? After all, once he was on his way, he could stay ahead of the storm fronts and out run them till he got to where they diverted in a different direction. After all, he'd been flying in the Southwest most of his adult life, and of the storms he remembered, most were fairly benign. Of course, most of those had been in the far west; but how much different could it be along the Gulf?

The Flight Back to Los Angeles — Day One

It was the morning of Les' third day of vacation when Flight Service informed him over the phone that there was a Severe Thunderstorm Watch in effect. Sigmet 'Alpha Two' was the designator for the storm. Now there are two kinds of storm warnings, Airmets and Sigmets. 'Airmets' are warnings to small airplanes to get out of the skies because of bad weather, while 'Sigmets' warn Boeing 747 sized airplanes to get out of the skies. The laws of Physics are not partial to size, causing destruction to all, if precautions are not heeded. Even the name 'Sigmet' (which stands for SIGnificant METeorological condition) should get a pilot's attention as something bad's about to happen, especially if it's in a pilot's flight path.

After the phone call, Les looked out the window. The sky was overcast and dark with a few breaks of sunlight shining down in bands, where there were narrow cracks between the swelling cumulus. It looked grim, but not terrible.

Finally, after thinking about it, he decided to make a run for it, realizing that he was a 'high time' pilot still alive after flying for most of his life. Besides, he only had eleven days vacation left, and if he didn't get started soon, he'd end up stranded somewhere short of Los Angeles. Besides, the overcast was at two thousand feet, and he thought that he could fly under it for a while, until clear of it. And since Florida is totally flat, what could possibly happen to him cruising along at a thousand feet?

It was 10 a.m. when Les said goodby to Bob Duffy, who wished him well and told him not to hesitate to turn around and come back, if the weather got too bad. Les thanked him for his hospitality, but was preoccupied with what Flight Service had said about the coming storm; it was slow moving and once here, would take a couple of days to clear out. The only problem was that behind that storm was another and another; each about three days apart, like a picket fence with spaces between them.

'After all, it's Spring. Wha' yawl expect?' the briefer chuckled over the phone.

Undaunted, Les climbed in the airplane to prepare for takeoff. All morning below the overcast the winds had been gentle, but now suddenly, as he was preflighting the Cadet, the wind's velocity increased, as well as becoming gusty. The storm's front was probably arriving. By the time he'd finished the preflight, several minutes later, the gusts were coming in waves.

With the winds picking up, Les started the Cadet's engine and performed an abbreviated engine run-up. He taxied to the end of the grass strip, turned into the wind and took off. As the Cadet climbed out over the runway, the wind gusts were strong enough to make the airplane's wings dip — first the right wing and then the left. Les fought the controls to keep the airplane level and steady.

From the ground, Bob Duffy, puffing on a large cigar, watched as his airplane left Florida forever. As it climbed out, it appeared that Les was rocking a 'thank you' and 'goodby' with it's wings. In the face of an approaching storm, Duffy thought that 'wing waving' was a rather 'cheeky' maneuver, considering it was like giving Mother Nature the 'finger.' And as Duffy stood watching until the Cadet was out of sight, he just shook his head, and then trudged towards the house before it started to rain, thinking that, before something terrible happened to Les, he'd better deposit the check.

●

Cruising along at twelve hundred feet, Les looked ahead at the darkening skies. The overcast was beginning to lower, and for the first time today he began to notice the humidity and began to sweat. The temperature wasn't high but beads of sweat appeared from each pore of his face and trickles ran down his checks and neck. The windshield was also affected by the humidity, and slowly it began fogging up. He turned on the Defroster and looked out the sides of the canopy; at least the sides were clear, allowing him to see forward a little.

For a long moment he allowed himself to enjoy the feeling of flying his very own airplane. It felt great! Someday he swore he would own his Mooney 21, but for now, the Cadet would do nicely. He settled into his seat, relaxing his shoulders, and tried to enjoy the moment, before he once again had to think about the weather.

He carefully moved the trim wheel forward a quarter of an inch, checking the airplane's stability by removing both hands from the control wheel. It was fine. Then he scanned the instrument panel and wondered if he could really out run thunderstorms at one hundred and ten miles per hour. He checked the airspeed indicator and thought that his real ground speed was probably twenty miles per hour slower, as the headwind increased. 'It's probably about eighty five,' he thought, but wasn't sure.

The skies were getting darker and he wasn't even at Charlotte Harbor yet. Duffy's estate was about twenty miles east of town and already the weather started to look threatening. He had the feeling it was going to be a long trip home.

Reaching into his flight bag on the seat next to him, he pulled out the Sectional chart and traced the route he'd planned weeks before. He traced the black line of his intended course on the map, with his index finger. It told him that his first leg back to California, from Duffy's to Tampa, was eighty miles away.

He started thinking that Florida may not have mountains rising up, but it surely had TV towers that could be as high as fifteen hundred feet, as well as six hundred foot radio towers. All of the guy wires secured the towers and stretched downward and outward maybe a quarter of a mile to the ground on all four sides. So, if flying into an unseen tower doesn't kill you, the guy wires will.

With his right thumb and index finger held two inches apart, he traced his courseline from Duffy's grass strip all the way to Tampa. The space between his thumb and finger accounted for ten miles of distance, either side of his course. His eyes scanned the route for the 'tall obstructions' symbol—suddenly finding one.

He glanced at his watch, then at his navigation log on the lap desk underneath the sectional. According to his calculations, he'd already passed whatever the chart's tall obstruction was.

'Jesus,' he sighed, peering around behind him as far as the canopy would allow. No tower in sight. A shiver went through him as he wondered how close he'd actually come to hitting it. But at the speed he was traveling, there was probably little or no chance for visual contact. So after a moment he resumed his gaze through the windshield, even though it was still partially fogged. It was apparent that the defroster was no match for this humidity.

His eyes searched through the clearer parts of the windshield, he could see the overcast of cumulus was sinking, and in another couple of miles he'd be in the 'soup' if he didn't descend below it.

His hand pushed the control wheel forward and held it until the airplane had descended to nine hundred, where he leveled off, instinctively he knew that this was a bad sign. The wisps of clouds and virga were now hanging from the dark and ragged bases of the overcast. The bases were still sinking, and forcing him to fly even lower. He backed off the throttle a hundred rpm and re-trimmed the airplane. Slowly, the Cadet began to bleed off altitude at a gradual rate.

Quickly, Les searched the sectional for the nearest airport, airfield, or crop dusters strip–anyplace where he could put down. Things were happening faster than he and his little airplane could handle. He wanted to get on the ground, and fast.

As his eyes searched the map, he remembered an old flying axiom. 'Its better to be on the ground, wishing you were in the air; than to be in the air, wishing you were on the ground.' And right now, he understood fully its meaning.

•

Suddenly with a thunderous roar, rain burst against the canopy, beginning with a sudden and startling deluge of water. The torrent came down with tremendous force, like from a fire hose. The airplane's speed against the headwind made the water's velocity over a hundred miles per hour. He was in the middle of a thunderstorm. The water hit the windshield so intensely that he lost all forward visibility.

He checked the altimeter as it was unwinding and about to reach eight hundred feet. He quickly leveled off and re-trimmed. He didn't dare go any lower.

The noise of the downpour became deafening — so loud and distracting that Les took off the tennis hat he wore when flying, and threw it behind him, reaching into his flight bag for his hard headset, quickly putting it on his head. Now, at least the storm's noise would be quieted. But, even with the headset covering his ears, he could still hear the staccato pounding of the rain on the canopy, although greatly attenuated.

Suddenly the wind gusts doubled their fury. The controls were moving in his hands even though he held the control wheel in a 'death grip.' The rudder pedals were alternately jabbing at first his right foot and then his left, as gusts of wind slapped the rudder's surface. It was now becoming very difficult to keep the airplane's attitude level.

He looked ahead through the sides of the canopy for a trace of lightning. You can't have a thunderstorm without lightning–that was a meteorological fact.

But, after gazing from wing tip to wing tip, out the canopy, no lightning. Maybe he wasn't in a thunderstorm; maybe it just felt like it.

So far, he hadn't bothered with the radio, since flying in VFR with plenty of visibility, he never really had to talk to anyone. And normally, he didn't like to be bothered chatting with Flight Service or Center. It just wasn't the way he'd been trained to fly. His idea of a pilot was someone who was self reliant and didn't need anybody on the ground to help him in the air. His plan for the flight back was to use only the built-in VOR capability of the Escort I radio. His flight plan called for landing at uncontrolled fields where thee wasn't any control tower to talk to, all the way back to California.

He wasn't sure if it was an omen, but the Cadet had come with a Push to talk switch, or PTT, allowing the pilot using a headset to use the radio without taking his hands off the control wheel. It connected between the radio and the headset and was velcroed to the control wheel. This design is far more convenient than having to pick up a separate mike all the time.

But, in this weather, it was no time to be timid about radio communications. He turned the volume knob up, and then flipped the mode switch from 'VOR' to 'COM.' He looked up the frequency for the Flight Service Station at St. Petersburg, on his navigation log, and dialed in the numbers.

'St. Petersburg Radio,' he yelled, into the boom mike of the headset, positioned in front of his lower lip. 'Cadet 4977 Lima,

Mooney Cadet, forty five miles south of you. Listening on 122.1. Over!'

With water cascading against the windshield, he was afraid that his voice wouldn't be heard over the background roar of the cloudburst. But after a long silence, a deep southern voice boomed in his headphones — a slow talking voice that was so deep and relaxed that it almost could put you at rest.

'Cahdit 4977 Limaah, thus hair's St. Petersbur' Radio. What kin ah duh fuh yawl?'

The grandfatherly voice sounded relaxed and confident.

'This is 77 Lima. I'm northwest bound at eight hundred, below an overcast, forty five south of you. In moderate turbulence. I need to land somewhere, and fast!'

'Rajjah,' the voice replied, slowly in a reassuring manner. 'How minny souls ahboard, an' how much fuel yawl have? Ovuh!'

'One soul aboard and three and a half hours of fuel.'

'Rajjah 77 Limaah. How's the weatha thair? Ovuh.'

'It raining so hard I can't see through the windshield. I'm indicating one hundred and five miles per hour, but I know my ground speed's a lot slower. I'm in moderate turbulence. And I'm unfamiliar with the area. I need to land at the nearest airport. Over!'

'Ah Rajjah that, 77 Limaah,' the voice returned, acknowledging.

There was a long silence, and Les' throat was getting dry and the pit of his stomach began to ache. For a pilot, that ache is the beginning of fear. And he knew that he had plenty of reason to be afraid.

'Ahh, 77 Limaah,' the voice began, returning suddenly. 'Aah'm lookin' at the weatha maps, an' aah think yawl's headin' is takin' yawl intah tha wors' part 'ah tha storm.'

There was another long pause.

'77 Limaah, ah sugges' yawl start headin' to 'bout 080 degrees. Thair' several small airfieds 'bout six miles from yawl's presen' locashun.'

Immediately Les banked the Cadet to the right, and watched as the compass in front of him slowly began to rotate on its axis. When the number 070 passed the lubber line on its face, he began to roll out, leveling the wings as the number 080 appeared.

A minute passed, and then another. Finally, Flight Service asked him if he could see any one of the airfields.

'Negative, St. Petersburg,' Les replied, his tone suddenly sounding urgent. 'Do you have a DF capability? I need a steer directly to the nearest airstrip. If you want, I'll declare an emergency!'

'Undahstan' 77 Limaah. Affirmative on the DF. Jus' key yawls mike for tin secons' without talkin' and wah'll git a fix on yawl. Ovuh.'

For what seemed an eternity, Les squeezed the push-to-talk button on the control wheel, squeezing it as if the tighter his grasp, the more power would be transmitted.

The rain was lessening slightly as he flew eastward, although it was still beating loudly on the canopy. But its tone was different; less intense. He wasn't sure what that meant, but any change at this point seemed good. When the ten seconds were up, he relaxed his grip.

The seconds went by. There was no reply. After a minute Les couldn't wait any longer and called back impatiently.

'St. Petersburg Radio? Are you still there? This is 77 Lima.'

'Rajjah 77 Limaah. Wee still triangulatin', Stan' bah.'

Another long pause. The rain was now actually lessening to a shower and now from the sides of the canopy he could see that the skies were several shades lighter towards the northeast.

How long does it take to do a DF fix. Les wondered? He

was worried that by the time they fixed his position, he'd already miss one of the airfields that he was flying towards. After all, in theory it was only supposed to take two stations to triangulate a position. But in this weather, anything could happen. The other station could be quite a distance away and the reception might be terrible in this storm. And of course, his altitude was too low for good reception.

A minute passed, and another. The skies were actually brightening a little, as the Cadet continued to the east. He wondered about his ground speed. If they didn't respond soon that six miles would be behind him. And he might even be flying past one of the strips even now!

Quickly he glanced out of each side of the canopy, trying to make out any kind of landing place, as the voice finally returned.

'Sorry fah tha delay, 77 Limaah. Got the fix. Show yawl 'bout four mahals from ah dustah strip. It'll look like ah durt road, but it's gaht fuel. Ovuh.'

Les stared at the sectional in his lap, as the voice described the airstrip.

'77 Limaah, the voice continued. 'Sugges' yawl descend to four hundrud, so yawl kin see bettah. It's commin' up at yawl's two o'clock possishun' and two mahals.

'Okay, St. Petersburg,' Les replied, sounding relieved and leaning forward to better see out of the right side of the canopy, as he throttled back to nineteen hundred rpm and pointed the nose down a 'bars width' below the horizon on the attitude indicator.

Slowly his eyes traveled from directly below the airplane to the horizon at his two o'clock position. Slowly, as if in a dream, on the horizon, out of the brightening skies, he could see a row of buildings begin to appear and as he got closer, he could see several airplanes parked wing to wing in a row near the buildings.

He banked the Cadet towards the distant buildings and

slowed down to pattern speed, and a minute later he could see that the parked airplanes were indeed crop dusters. Two Pawnees, a Thrush and three Agwagons.

'St. Petersburg, this is 77 Lima. I've got the field in sight! he yelled into the mike, banking the airplane in a wide arc, as he over-flew the area looking for power lines. 'Thanks for your help!'

'Okay 77 Limaah,' the voice replied, sounding relieved. 'Whin yawl's on the groun' give us a call un tha lan' line, so we know yawl made it. Hair?'

'WILCO, Out.'

As the Cadet circled the strip, the rain on the windshield lessened to a drizzle. Les could actually see forward through the windshield again. He flew a wide downwind leg and then a long final approach, as the Cadet descended to two hundred feet, where his forward visibility greatly improved for the first time since the rain began.

He pulled on the carburetor heat and began easing the throttle back to twelve hundred rpm as the Cadet approached the beginning of the airstrip. Les pulled the flap lever for full flaps and then chopped the throttle back to idle when the landing was assured.

The airplane leveled off and slowly mushed onto the hard clay of the airstrip, bouncing twice before settling down. The Cadet took the entire thousand feet of the dirt strip on rollout, coming to a stop at the very end.

Les sat there without moving in the Cadet at the end of the dirt strip, the engine idling. Sweat trickled down the sides of his head and he noticed that his shirt was damp and limp. He opened the canopy several inches so the prop wash would cool him. His body was one tight knot and drained of energy. And as he sat there, he thought about how close he'd come to disaster.

Waves of emotion swept over him as the dangers of what

could've happened began to reach his understanding, his hands trembled and he shook his head wondering how he could've allowed himself to get into such a foolhardy situation.

After a long moment, he retracted the flaps and gunned the engine enough to turn the airplane around and taxi back towards the buildings at the other end of the strip. And as he taxied he thought about how he was going to get the airplane to California in weather like this.

•

This was the end of the first day of his trip back to California, and he realized two things: first, that he'd never encountered weather quite like this before, and second, after only two hours of flying, he was just fifty miles farther north from where he'd taken off. At this rate, he thought, if the weather didn't improve, his vacation would be over and he'd still be somewhere in Florida.

The Flight Back to Los Angeles — Day Two

The second day began pretty much like the first, and the call to Flight Service from the motel sounded like a recording of the previous day's weather. It had rained during the night, but now it was down to just intermittent drizzle, and the wind was calm.

Les was getting concerned about what remained of his vacation. He'd already used up three days with very little progress. In fact, his worry now was about the weather and having to leave his airplane somewhere short of California. The thought of not making it back began to depress him, and motivated him even more to press on.

Flight Service had told him about the storms in Spring and about their frequency, arriving at a rate of one every four days or so, with spaces in between each of a day or two. Sometimes they arrive one after another with no breaks–just moving right on through; while other times, they would slow down and wouldn't move for several days. When the storms were stationary for several days there was nothing to look forward to but continuous rain. When that happened, there was nothing to do but wait it out. And since Meteorology was still mostly an art, the storm's speed could not really be predicted accurately.

Les wanted to get home, and he needed to make some progress every day. He thought that if he could only get to Midland, then he'd be out of the storm track. The remainder of the trip would be in VFR weather.

At breakfast that morning, in the coffee shop of the motel where he'd spent the night, he took his Sectional. As he sat studying his routing and filling his stomach with biscuits and gravy, two middle-aged men drove up in a brown pickup truck and parked in front of the window near his table. He watched as they got out into the morning drizzle.

The sign on the door of the truck read 'Thomas Brothers Crop Dusters.' Below the name was a cartoon of a biplane.

The two men stomped their feet and shook the drizzle off their clothes and waited for the waitress to seat them.

'Thar's no way we're gonnah flah tahday, dammit to hail!' the taller of the two said, wearing a blue baseball cap and a faded black tee shirt.

A waitress approached and led them to a nearby table.

'Ah know,' the other replied, shaking his head, as they sat down. He looked like the Marlboro Man–a lanky body and a lean and weathered face, wearing a white straw cowboy hat square on his head, and a white tee shirt in stark contrast to his dark sun-tan.

As they opened their menus, the man in the cowboy hat spotted Les studying his sectional and began laughing. He nudged his partner's arm with his hand.

That guy mus' thank he's a goin' flahun' tahday,' he chuckled. 'Hey fellah,' he laughed. 'Goin' somewhere's in this hair weathah?'

Les looked up and smiled. 'You pilots?' he asked.

'Dustahs,' the one in the cap replied. 'Agwagons and Thrushs mostly, and once in ah while a Pawnee, fur Thomas Brothers. Yawl from aroun' hair?'

'California. Los Angeles, really. I'm the one that landed at your strip yesterday afternoon in the storm. That Cadet's mine. Bought it in Fort Meyers and I'm trying to get it back to Los Angeles.'

They all laughed.

'That Caydet's a nice little plain' the man in the baseball cap observed.

'It's small, but it's all mine,' Les replied.

'Well, yawl nevah git too fah in this hair weatha usin' secshunals. Yawl got to use the rat kindah maps.'

'What do ya mean?' Les frowned.

'Wall,' the man in the cowboy hat said, chiming in. 'Whair thair's nothun' to see but roads when yawl lookin' down in thaht rain, roadmap's thah only thangs that'll help yawl.'

'Really?' Les remarked, surprised. 'But, that makes sense.'

The waitress came over and poured two coffees for the duster pilots, and then returned to the counter.

'Thah' only way tuh flah in thus hair weathah,' the one in the cowboy hat continued, 'is by lookin' down and followin' roads.'

'Thah's fur dam shur! the other added.

And with that, Les finished his breakfast and bought a roadmap of Florida at the motel office as he checked out. He called a cab to take him to the duster strip and was determined to get going no matter what. Especially since he was now armed with a secret weapon for stormy weather navigation.

It was another hour before he could get into the air, during a pause in the drizzle that lured him onward.

•

He thanked everyone at Thomas Brothers that had been so helpful during his overnight stay, and for letting him park his airplane in their hangar. He tried to pay for a night's parking, but no one would take his money. I guess they thought that his flight was doomed and he was destined to die, while tempting the weather gods. They also probably thought that if they took his money, his death would be on their heads.

With a drizzle still falling and obscuring his forward vision, Les opted to make a dash for it, and executed a 'soft field take-off' on the slick wet clay of the airstrip. Full power, control wheel all the way back, with a notch of flaps. After several seconds at full power the Cadet sloshed along through the puddles of standing water finally mushing into the sky just a few feet above the ground. Les lowered the nose to level off and pick up airspeed.

As the airplane finally reached flying speed, he raised its nose and quickly glanced over at the open hangar, where the duster pilots were all grouped out of the weather, watching his departure. He waved as the Cadet passed them at eighty miles per hour, but as a group, they looked like they were watching the launching of the Titanic. No one waved back; they just stared, as the small airplane staggered into the sky disappearing after a long moment, below the distant and brooding clouds.

Les initially climbed up to fifteen hundred and leveled off, then turned slowly to the left and took up a heading of 265 degrees, a heading that would intersect with one of the highways that Les marked on the roadmap, as his course.

The skies were darkening as he headed west, but the horizontal inflight visibility was at least two miles. And from his altitude, he could at least follow the roads fairly well. And, after a few minutes of pilotage, he intercepted the county road and followed it to the interstate. Once over the grass median of the interstate, he decided to use it as his course, and began tracking the median.

With something tangible to watch below, Les began feeling comfortable, since the interstate would lead him to where he wanted to go.

Once firmly established over the interstate, Les called Flight Service on the radio to offer a Pilot Report.

St. Petersburg Radio, Cadet 4977 Lima, transmitting on 122.1.'

There was a long silence and when Flight Service finally responded, it was with the voice of a 'Redondo Beach surfer,' not a southerner, like the day before.

'Roger, 4977 Lima. This is St. Petersburg. You're number two, back in a minute.'

When Les heard the voice he laughed out loud. He expected a southern drawl, not a voice from California, and for an instant he couldn't understand what was going on. Then he realized that Civil Service kept its personnel homogenized, by rotating them every few years.

When Flight Service returned, the voice took Les' Pilot Report and read him a report from an airplane fifty miles further north, somewhere around Tallahassee. The pilot reported that the weather seemed to be getting worse; about like what Les experienced the day before.

It turned out that there was an Airmet in effect for the Gulf coast from St. Petersburg to New Orleans; pretty much Les' intended route, and his newest problem now was a fast moving storm from the southwest that could overtake him. He needed to know just how fast it was moving so he could plan to land somewhere prior to its arrival.

He copied the report in pencil on his lap board, so he could compare it with later reports, to see trends developing, if any. And although the pilot fifty miles up the line gave a fairly complete report, the storm's speed was still an unknown, other than it was fast moving.

The Cadet's indicated airspeed was one hundred miles per hour, but as Les looked at the ground out the side of the canopy, it didn't seem to be that fast. The wind velocity and direction weren't clear to him. But, he realized that he had to hold a three degree crab angle to the left, just to keep centered over the interstate.

St. Petersburg, 77 Lima,' he said, squeezing the 'push to talk' button. 'I'm estimating a ground speed of about eighty miles per hour, over!'

'Roger, 77 Lima.' the 'surfer' acknowledged. 'At your present ground speed, the front should be to Gainesville by the time you get there–and you'll never make it to Tallahassee. You'll be lucky to make it to Cross City.'

Les heard the words, and checked the road map. It would take him at least two hours to make Cross City from where he was.

'Roger, St. Petersburg,' he acknowledged, his voice sounding irritated. 'I'm trying to get this airplane back to California by the end of the week. I think I'll keep going towards Tallahassee and see how far I can get before the storm sets in.'

'Roger, 77 Lima. But don't wait too long. What's your current ceiling?'

'It's hard to tell, but I think about three thousand. It looks lower to the northwest.'

'Roger 77 Lima. Be apprised that there is now actually two storm fronts; one faster than the other. One is warmer and is being overtaken by the faster cold front. Sometime tonight they'll probably occlude and by tomorrow–it'll being raining continually for forty days and forty nights,' he laughed.

'Thanks for the good news,' Les laughed. He wondered just how long it really would rain, if the occluded front went stationary.

The airplane was flying well and felt good. The engine sounded strong and smooth, but as the minutes passed, Les had to increase the Cadet's crab angle a few more degrees, which meant that the wind's direction was changing, as he approached St. Petersburg from the southwest. The visibility was still clear enough that he could see beyond the city, where the ceiling was much lower.

He looked at his watch, then calculated his ground speed. It was fourteen miles every ten minutes.

The first two fingers of his right hand walked up the map an inch at a time; each inch equaled fourteen miles or ten minutes. It didn't look good. At this rate, he'd only be abeam Ocala when he'd meet the storm. But Ocala was thirty miles south of Gainesville. His finger traced a line due west of Ocala. It looked like he'd have to put down on the coast at Cross City. Maybe even sooner, if the quartering headwind got any stronger. He started searching his sectional for airports. The pickings were slim.

As he scanned the chart, a 'tall obstruction symbol' caught his eye. It was either a commercial radio antenna or a TV antenna tower. And it was right on the interstate highway he was tracking. It was twenty miles this side of Cross City. He took out his pencil and circled it. Then he looked at the road map and drew it in approximately where it was should be geographically.

He marked an 'X' where the tower should be. He hoped his

approximation was fairly accurate, because when the time came, if he was too low, then that's where his trip would end.

•

The ceiling was getting lower and he didn't want to get caught in a cloud, so he trimmed the airplane to just a hair of nose down attitude, then watched the vertical speed indicator read a descent rate of two hundred feet per minute. That was probably too fast, but it'd be okay for a short while. His eyes moved from the vertical speed indictor to the altimeter, and at one thousand feet, he raised the nose and re-trimmed for level flight. A moment later the vertical speed indicator verified the attitude with the pointer again resting on 'O.'

He glanced out at the side of the canopy just ahead of the leading edge of the left wing. One hundred and ninety miles away, Tallahassee was waiting somewhere beyond the horizon. But as hard as he stared in that direction, there was nothing to see but darkness. The storm was coming for sure.

Les tuned in Gainesville Radio and listened for any conversation with pilots who might be flying towards Tallahassee. Nothing. And after several minutes of silence, it was pretty obvious that he was the only flyer dumb enough to be still milling around the sky during an Airmet. He waited several more minutes and when the frequency was still quiet, he gave Flight Service a call.

A long minute went by and then another. He knew that the radio was working. He couldn't understand why nobody answered. He gave them another call. It was a good twenty seconds before a loud voice boomed in his headphones.

'Aircraft callin' Gainesville Radio, yawl very weak. Say agin' yawl's call sign.'

It was another southerner and he took it as a good omen.

'This is Mooney Cadet 4977 Lima.'

Suddenly Les realized why Gainesville hadn't answered right away. It was his altitude. Not only was he more than eighty miles away, but the important thing was that his altitude was only one thousand feet. He smiled to himself. He was lucky anybody heard him at all.

'Rojjah, undahstan' yawl's 4977 Limaah. Say yawl's pussishun.'

'77 Lima is twenty north of St. Petersburg at one thousand on a VFR flight plan from Fort Meyers to Tallahassee.'

'Rojjah, say yawl's requess.'

'Are there any recent PIREPs from the Tallahassee area. Over.'

'Affirmative, 77 Limaah. Jus' talkin' to aah Bonanza over Greenville VOR, 'bout thirty five northess' of Tallahassee. He's in IMC at three thousan' five hundrid with the win' gustin' to thirty an' rain.'

'Roger, Gainesville. What are the ceilings at Tallahassee.'

'77 Limaah, Tallahassee's IMC. Ceilin's three hundrid an' aah RVR of twenty four hundrid in fog an' rain. Say yawl's intenshuns.'

'I'd like to get to Cross City VFR if I could. Over.'

'Yawl betta hurry, 77 Limaah. Storm's ah commin' fass.'

'Roger, Gainesville. I'll keep guarding your frequency, and let ya know when the weather gets too marginal. Over.'

'Aah rat 77 Limaah. Yawl fluh safe, hair?'

'Thanks for your help. 77 Lima Out.'

•

Les studied the sectional in his lap and calculated his present position to be about eighty miles south of Cross City. If there wasn't any wind, it would take him about forty five minutes to get there, but there was a lot of wind and pretty gusty, too. The quartering winds were really picking up in intensity, and the three degree crab angle went to five degrees, and now he was holding eight degrees. Not only was the wind increasing, but as always, it was a headwind.

With a fast moving storm he could forget about making Cross City. His ground speed would really be reduced. He'd have to put down somewhere south of there.

Ten minutes went by and he was now about ten miles further north. The ceiling was really lowering now and he couldn't see more than a mile forward. He had to get below the clouds and stay there.

His right thumb and index finger pinched the trim wheel, pressuring the wheel imperceptibly and moving it just enough forward. And after several seconds the pointer of the vertical speed indicator moved down very slightly to register fifty feet per minute of descent. And then after several more seconds, the long pointer of the altimeter began to unwind, as the Cadet's altitude slowly bled off.

As he overflew Crystal River, fifty five miles south of Cross City, his altitude was down to seven hundred feet — way below VFR minimums. The problem was that even as low as that, there was still a heavy mist forming on his windshield and thickening until the windshield became blurry. He was barely below the overcast, but virga was falling–the front was only minutes away.

He decided not to bother Flight Service anymore, especially after what happened yesterday. Although they could help him find places to land, it just took too long. He thought he was better off on his own.

Forward visibility was becoming a problem now. The mist that had glazed the windshield was now turning to rain and the airstream against the plexiglas pushed the rain drops upwards in rivulets from the windshield's base. There wasn't any sudden deluge like the other day, but the outcroppings of the storm was here and had to be reckoned with.

The altimeter was still unwinding as he passed through four hundred feet. He could see the interstate very well when he looked straight down from the sides of the canopy. He just couldn't see forward anymore.

He looked at his sectional again and then at the roadmap. There was a small private airfield southeast of Cross City. It was called 'White's Farm.' It was a private airstrip, but this was a kind of emergency. Besides, it was probably more improved than the duster strip he'd landed on yesterday. And for an instant he wondered about its name. If it was a farm, what did they grow? He decided that it must be some kind of horse ranch.

How long could it take to get there? He took out the circular flight computer from his shirt pocket, setting up the movable scales with his right hand. He looked under the curser. The answer was twenty two minutes from his present position. Did he have twenty two minutes? It was too long. And looking at the maps again told him that there was nowhere else to land except on the interstate itself.

In a life threatening situation, landing on the interstate as an emergency measure was always Plan A. But if the law in Florida was the same as in Georgia, then anyone who landed on a state highway was required to disassemble the airplane's wings and truck the airplane to the nearest airport–all at the pilot's expense, where it would have to be reassembled and rerigged prior to flight. It wasn't like California, where they'd let you fly it off the freeway just to get it out of there as an obstacle to traffic. Needless to say, landing on roads wasn't an option to be exercised unless it was 'life or death.'

Now as he recalculated his ground speed, he realized that the storm would probably catch him even before he could get to White's Farm. His ground speed was now down to less than seventy miles per hour.

The wind was gusting now and he could feel each wave of air as it smashed against the sides of the Cadet, like ocean waves against a small boat.

And his feet could feel each bounce of the rudder pedals as the gusts hit the rudder broadside, trying to move it.

The sky was so dark now that to see the instruments clearly, Les turned on the overhead night vision red flood light that illuminated the instrument panel. He turned the rheostat up to 'full bright.' The white numbers on the indicators of the panel took on a pink glow.

Just three hundred feet below him on the interstate, four cars that had braved the weather were driving, one behind the other. They were probably going about sixty in the diminishing visibility. They were evenly spaced and kept a constant distance between them. Their speed and following distance was probably a function of the speed of their windsheild wipers, and how long between cycles their view of the road was obstructed.

Suddenly, Les realized that the four cars were going just as fast as the Cadet. That meant that his ground speed was now down to about a mile per minute. As he stared at the cars, their tops glistened in the pelting rain, which now was coming down with enough force to be heard against the Cadet's canopy.

Then without warning, the wind shifted direction. The storm front was here. And instantly the eight degrees of crab of left rudder that'd kept the Cadet centered over the interstate, changed to eight degrees of right rudder.

As soon as Les realized what was happening, he tightened his seat belt as well as his grip on the control wheel, banking the airplane to the right to keep it centered above the median. And with the storm's front upon him, the benign gusts against the Cadet's hull that had been just an aggravation, now became serious turbulence, causing the control wheel to be jerked out of Les' grip. As each gust thumped the underside of the wings, the control wheel jerked as the ailerons were abruptly deflected by the winds force.

Les glanced at the altimeter and knew he was too low. If he was ever caught flying this low the Feds would 'write him up' for sure; but this was an emergency. He couldn't really go any higher or he'd be in the clouds. And as he strained to see through the rivulets of water in front of him, it was then that he remembered about the tall obstruction near Cross City. The radio or TV tower–whatever it was. If it was a TV tower, it's height could be hidden in the clouds. If it was a radio broadcast tower, it's height was probably only four or five hundred feet. But either way, now that the Cadet was mushing along at three hundred and still descending to keep under the sinking clouds, if he collided with it, it would happen so fast that he'd never know it.

If he was able to see any distance ahead, it wouldn't be so bad, but with less than a quarter of a mile forward through the blurry windshield there was no way he'd ever see the tower in time.

And now the ache of anxiety that began in the pit of his stomach had grown to the jabbing pain of fear.

When the Cadet descended to one hundred and fifty feet by the altimeter, Les leveled off. At that altitude, with the average elevation of the interstate at about one hundred feet, put the Cadet at about fifty feet above the ground. He absolutely couldn't go any lower. He was as high now as the power lines. This situation was now life and death. At this altitude and speed, one false move would cost him his life.

He trimmed the airplane for level flight and looked at his watch. He still had fifteen miles to go. Another quarter of an hour of flying and it still wasn't for sure that he'd arrive where he wanted to. He really wasn't sure of exactly where he was–only that he was on the interstate.

Outside the cockpit the visibility was fading and a fog was forming in the midst of the rain showers. With visibility reduced to just a few hundred yards, Les had to rely on just tracking the median of the interstate, while staring straight down from the sides of the canopy, as forward visibility was now out of the question with darkness settling in.

Trying to make Cross City now was just a 'wish.' There was nothing to do, but to wait until the time ran out and hope to see 'White's Farm' off to the right. And if the time ran out and he couldn't find White's Farm within several minutes, then he'd have to land on the interstate and suffer the consequences. He should've put down before the weather worsened, somewhere around Crystal River or Inverness.

And now, here he was again, flying at an altitude of fifty feet above the ground with a ground speed of probably around sixty miles per hour. Only somebody with a death wish would ever consider flying like this. And since he couldn't see out of the windshield anymore, if the tower was closing on him, He'd never see it in enough time to react. Death would be instantaneous.Not a bad way to die, he thought, if one wanted to. But, his problem was, that he was trying to stay alive and survive. That's always a lot harder to do than giving up and dying.

His face was frozen in a frown just thinking about his situation. After all, he had over ten thousand flight hours in his log books, and here he was, for the second day in a row,'stuck in a crack,' flying in bad weather, like a student pilot, and totally without any options left, except to continue flying in hopes of spotting some farmer's airstrip. Some CFI he was.

The storm was worsening by the minute, and he glanced at his watch, wishing that the sweep second hand would move faster. But there were still six minutes to go. He hoped his dead reckoning was accurate, otherwise he'd only have a couple of minutes to circle around looking for the airstrip. When the time ran out he'd have to land–it was just raining too hard to go on.

Les peered intently through the rivulets tracing their way upwards on the windshield. He couldn't see beyond them anymore, but psychologically it made him feel good to pretend that he could.

Suddenly something faintly red flashed once through the blurry rivulets of the windshield and then was past. It flashed by, off the left wing tip of the Cadet. It happened too fast to identify or react to. Les only saw it flash once and was gone.

Had he imagined it? Was it real? He was very tired and maybe it was just his mind hallucinating, knowing that he very much was looking out for a 'flashing something.'

The jabbing pain in his stomach suddenly went away. He was drained emotionally and was so tired that he couldn't react reflexively. Whatever happened now was not only beyond his 'feeling,' but beyond his control as well. He had crossed over from being a pilot in his own airplane, to a passenger just along for the ride. He'd been overtaken by events beyond his control and was just waiting now for the seconds to count down.

The situation was ludicrous, really. Here he was in a tiny metal airplane with a wingspan of thirty feet, with a tiny ninety horse engine, trying to fly through a storm with winds so high, that cars were passing him as he flew over the interstate.

His dream flight back to California that he'd so carefully planned had turned into a nightmare. This was the kind of horror story you read about in flying magazines under the heading of 'Never Again.'

His eyes were transfixed on the sweep second hand of his watch. The hand slowly traced an arc around all the points of a circle, oblivious to his plight. The watch cared little about his outcome, whether it would be life or death.

The skies were now almost as dark as night, as the time finally expired. He gazed out of the right side of the canopy. But in the fading light, would he be able to see anything? A moment later something off the right wing tip came into view at three o' clock. It caught his eye. Through the slanting rain he could see a collection of four white buildings a quarter of a mile in from the shoulder of the interstate, beyond a dozen acres of pastures.

He immediately slowed the Cadet down to maneuvering speed and banked towards the buildings. Even in the rain he could see a dark paved airstrip beyond several barns and behind an exercise track for horses.

The wet black asphalt of an airstrip glistening in the rain stood out in relief against a more subdued background. Even the white painted threshold marking and painted side lines of the runway were visible through the rain.

His face broke out in a wide smile at the sight, as the painted words 'WHITE'S FARM' appeared on the runway in ten foot letters.

•

After he landed and rolled out to the end of the runway, he sat in the Cadet for a long time with the rain drumming on the canopy and the engine idling. It wasn't until then that he realized that his entire shirt was soaking wet.

'Godammit!' he yelled out loud, taking off his headset and

rubbing his ears. 'How could I have been so stupid?' Getting back to California wasn't worth dying. It wasn't life and death. But, twice in two days his flights had ended in life and death emergencies. Getting his airplane safely to the West Coast wasn't worth his life. What was driving him so hard?

He sat there for several more minutes trying to calm down, and then throttled up and taxied back to what looked like a hangar near the other end of the parallel taxiway.

So, with the end of Day Two of his trip back, he'd only covered a total of one hundred and forty eight miles, with another day of extreme risk to life and limb. When would this insanity end? Two days and two hundred and two miles completed, out of a total of twenty three hundred.

He felt very tired. He was too tired to think about things right now. What he needed was a good night's sleep and something to eat. He looked at his watch. It was only 1 p.m. As he trudged through the rain shower towards what looked like the main house among the cluster of buildings, he wondered if his Mooney Cadet was really worth going through hell, for? It was a question that had no answer.

•

It wasn't until the next morning, when Les was talking to one of the farm hands that he was able to identify the faint flash of red that he'd passed, while flying over the interstate. It seems that there were two towers — one on each side of the interstate, opposite each other. He must've passed directly between them as he flew over the interstate's median. He didn't think much about it, until the farm hand told him that he was more than lucky to be alive.

It turned out that the two towers were connected for stability, by a two-inch thick cable, that stretched across the interstate at an elevation of one hundred feet above the ground. Les suddenly realized that he must've flown under the cable, missing it by just fifty feet! Escaping death in the midst of a storm by the mere luck of fifty feet–somebody 'up there' must really like him a lot. Coming that close to death and surviving, doesn't happen to a pilot twice in one lifetime!

As the reality of the towers settled in to Les' understanding, he suddenly felt very old, and his knees began to tremble.

•

Day Three through Day Nine

Day Three was a better travel day than day two, as the storm passed through the Florida panhandle and on towards the Atlantic. On the third day, Les was able to fly out of Florida, from White's Farm to Crestview, Alabama, a distance of two hundred and forty miles.

Day Four was spent on the ground, as well as Day Five — 'wishing he was in the air.' The only good thing about being grounded in Crestview was that the town was famous for its seafood. And since Les loved seafood, he didn't mind too much that he was stuck there for a while. And even though it rained steadily the two days that he was 'grounded,' he told me later that the place was very tolerable, because the seafood was fabulous.

Day Six was another marginal VFR day, but he was able to takeoff and cover another two hundred and seventy miles from Crestview to Picyunne, Mississippi. The flight was the usual rain showers and fog and then breaks of sunshine for ten minutes and then rain again. It happened so regularly that Les was almost getting use to it.

The weather on Day Seven wasn't as good as Day Six, but Les still covered one hundred and eighty miles — from Picyunne to Eunice, Texas.

By the end of Day Eight, Les had traveled from Eunice to Hawthorne, Texas — a distance of one hundred and ten miles. It was still slow going as he was still caught in the storm track.

On Day Nine, he flew from Hawthorne, Texas to Cleveland, Texas in very heavy rain — as heavy as the weather had been on Day One. The total distance covered on Day Nine was just forty miles.

Day Ten

During the week while Les was flying back from Florida he called me at home each night to let me know of his progress, as well as the horror stories about the weather. The way the weather sounded to me, I wasn't sure that he was going to be able to meet me in Austin on Thursday, as planned. Although the weather was a factor, he was fairly certain that he could be there by then.

Since he sounded so confident on the phone, I caught the 'red eye' to Austin from Los Angeles at midnight, Wednesday.

My flight was on Trans Texas Airlines, an airline that has since gone under. It was a shame, but it really couldn't survive for very long. The employees made sure that it was fun to fly on their carrier. The service was terrific and the ambiance was informal. The price of a ticket on Trans Texas to anywhere was the lowest of any scheduled airline in the world. It reminded me of PSA, Piedmont and Frontier and a couple of others from the old days–before the trunk carriers like US Air gobbled up all of the little ones and turned commercial aviation into a marathon of misery for the airborne traveler.

To give you an idea of how much fun working for Trans Texas was, the uniforms for the stews were white Tee shirts with the Trans Texas logo on the front and khaki shorts. It sure was a pleasure to fly in such an informal atmosphere. But those days are long since gone, never to return again.

Anyway, the flight departed at midnight from Los Angeles with an hour's layover in Las Vegas. After taking off from Las Vegas, the flight was then non-stop to Austin. The night skies were very dark, once we got to El Paso, where we began to encounter thunderstorms. As we flew over El Paso, thunderstorms were visible in the night sky for hundreds of miles in all directions. That, of course, should've been a clue to me or at least an omen, that Les could hardly guarantee arriving at the Executive Airport on time–with all of that bad weather to fly through.

But, watching the lightning show, as it streaked from cloud to cloud and then from cloud to ground, looked to me from thirty five thousand feet as pure entertainment. It looked like the entire state of Texas was under siege by the storm gods. And I don't know why, but I never really considered that the weather I was watching was the same weather Les had to fly through. And that there was a very good chance that it could be a factor in his ability to make any kind of progress. All I could think about was the fun I was going to have flying his airplane back to Los Angeles.

•

By the time the airliner landed in Austin, it was 5:30 a.m., and it wasn't raining anymore. The glow of 'first light' was back lighting various layers of the overcast that was fast turning into scattered clouds. It was still so early that even the airport coffee shop wasn't going to open for another hour and a half.

For some reason, instead of relaxing until I could get a hardy breakfast, I was impatient to get to the Executive Airport, figuring that I could always eat breakfast there. This strategy would prove to be a big mistake.

It's amazing how when a person wants something to happen so badly that they focus on nothing else, reason and logic seem to disappear and their vision shrinks down to a narrow tunnel. And that's exactly what happened to me. With bad weather totally covering the state of Texas, logic would dictate that there was no way that Les, flying his little airplane through one thunderstorm after another, could possibly arrive at the Executive Airport at the appointed time. If I'd been thinking clearly, for me to rush right over there was silly. There would be plenty of time for me to get there by noon, after all, it was only ten miles away.

But, I was so impatient to get out of Los Angeles and begin an adventure. I completely disregarded all the obvious facts, focusing only on the fun that would begin when Les arrived.

My first problem at Austin — not finding breakfast — was forgotten, now that my second problem of finding a way to get to the Executive Airport confronted me. Finding a cab driver that knew where the Executive Airport was, and then a freeway routing that would get him there, was suddenly taking on great importance. The fact that two out of two cab drivers had never heard of the place should've been a 'sign. If cabbies didn't know about it, maybe not too many people ever went there. And, if nobody went there, what is the likelihood of a coffee shop or restaurant on the field? But, undaunted, I questioned several more drivers, and pressed on with my quest.

I finally settled on one cabby who told me that although he'd never been there himself, he knew exactly where it was and had plenty of maps to show us how to get there. By this time I was too tired to argue and just got in his cab.

Although nothing was working out the way I'd planned, in retrospect I know that I would have been a whole lot better off if I'd just waited and had breakfast at the Airport coffee shop. The flight had been long and I couldn't sleep on the airplane, and hadn't really slept since the day before yesterday. And now I was hungry and very tired and my mind was not at all up to thinking.

The skies were clearing as the sun came up, and I actually believed that Les would fly right in at noon, as advertised, regardless of what horrendous storms lay between us.

As the cab sped along the freeway, I began to have second thoughts about this adventure that I'd planned. The fun excursion was growing into something that was anything but fun. Its scope was growing larger than what I'd ever bargained for. It was becoming a giant mistake on my part. And the greatest shock of all to my system, in my somewhat 'degraded' state, was that the fantasy of the trip so far, in no way was at all close to it's reality.

By the time the cabby found the Executive Airport, the sun was actually shining brightly and the clouds were all but gone from the morning sky. And as he dropped me off at the entrance to the access road to the airfield, I realized that the view in the distance of the airport was not at all what I expected a metropolitan executive airport to look like.

I had envisioned something more like a municipal sized controlled airport like the many executive airports around large cities in California, where Citations and King Airs flew in an out by the hour and where the facilities of corporate pilot lounges rivaled those of a Hilton Hotel. Instead, what I was looking at was an uncontrolled country airport. It did have a paved and lighted runway, and a modest attempt at an industrial park of five separate cement buildings at the far end of the field.

Amongst the collection of buildings was one that looked

like an airport's terminal building. But, as my eyes scanned the distant complex, the one thing I was looking for and couldn't see were signs announcing a restaurant or coffee shop.

•

As I trudged down the gravel road towards the airfield, the thought occurred to me that maybe I wasn't even at the right airport. What if this wasn't the Executive Airport and Les would fly into the real one, while I was waiting for him here. He wouldn't see me. He'd refuel and leave, per our agreement.

As I walked in the chilly morning air, the thought that I was in the wrong place quickly passed. But one thought lingered as a chill went through me, and that was the fact that although I considered myself a seasoned traveler, the few things I packed and brought along were totally inadequate for this place. For some reason I didn't bring a heavy coat or jacket. All I had on was a sweatshirt under a light wind breaker, which was much too light to warm me against the morning temperatures of the low forties. As a naïve traveler, I assumed that the April temperatures in Texas were the same as those in Los Angeles. What a 'rube' I really was.

•

I trudged onward through a large field in ankle high weeds still wet from last night's rain. Not only was I tired and hungry, but I was also cold. And no matter how much I moved around I wasn't going to get any warmer. I don't know what I was thinking about when I packed for this trip. I sure didn't come prepared to stay for more than a few hours. My fantasy about this trip didn't allow more than a few hours of standing at the airfield waiting for Les. And I didn't bring more than twenty five dollars in cash, and a master card.

In the distance near the edge of the field was a truck and four people standing in a group, staring at something in front of them. The group turned out to be three young women in their early twenties, and a man of the same age. What they were looking at, at their feet was a folded up hot air balloon.

'Hello,' one of the women shouted, as they all turned to look in my direction. 'Come on over.'

Surprised, I looked behind me to see whom they were hailing, as I wasn't used to girls beckoning me. But when I looked behind, there was no one else. They were calling me. I couldn't imagine why a group of women wanted me to join them. But, having nothing else in particular to do until noon, I trudged over to them looking like Willy Loman from 'Death of a Salesman,' head down, carrying a tote bag in each hand.

As I got closer, it was apparent that the truck had transported all of the people and the equipment that was unloaded on the ground.

'Good morning,' I said. 'What's going on?'

'We're about to inflate a hot air balloon,' the young man offered. 'Ever seen one before?'

'Not up close.'

'We could use another pair of hands, if you'd like to help,' one of the women said.

'Sure. Just tell me what to do.'

And with that, they all started to unpack the equipment. I stood back out of the way and watched as two of them unrolled the balloon itself. It was interesting to see what was involved in setting up a balloon launch. The one piece of equipment I didn't expect to see was an electric window fan, one of those square box shaped things.

One of the women saw me staring at it remarked that it was

for filling the balloon with ambient air, to increase the bag's volume prior to filling it with hot air from the butane heater.

I nodded, as if I understood, and continued to watch as they finally got things organized. As they were unpacking I heard them talking to one another, and noticed that none of them seemed to have a Texas accent. They all talked like Californians. I found out later from a cab driver that since Austin is a university town, everyone there is mostly from somewhere else. It turns out that to hear people talking with a drawl, you have to visit one of the smaller towns down the road. Even Dallas is too big to sound regional. The people there talk like they're from Los Angeles.

When the balloon's ground crew were ready for me to help, my job was to stand by the open mouth of the balloon's bag and hold the window fan in an upright position, while it forced air into the bag. A short distance away, a gasoline electric generator was started. Fortunately for me, the sound of the generator broke the morning silence shortly before I could ask the question of where were they going to get electricity for the fan?

While my job was stationary, the women scurried about, staking out tether lines and whatever else they were supposed to do, while the young man knelt on the ground near me, checking out the plumbing on the butane heaters and the tanks that were strapped to each corner of the balloon's square wicker basket.

He connected the top of the aluminum frame to the balloon itself. The heater jet assembly was now horizontal to the earth, and connected to the mouth of the balloon. The balloon's bag was now filled to a third of its total volume as the man ignited the butane burners. Soon hot air was shooting into the mouth of the bag.

At this point my job was completed, as the balloon slowly began to erect itself. And after another twenty minutes, the bag was standing four stories tall, and tugging at its lines. When it was completely inflated, I asked one of the women just who all was going to ride in it.

'Oh Jerry is. It's his balloon. We just come along to help set it up.'

'Do ya have to follow him in the truck, to pick him up later, after he comes down?'

'Oh sure,' she laughed. 'We get to follow as close as the roads will allow, and when it comes down, we go and find him and help pack it all up.'

'What happens if ya lose sight of him.'

'We use the two way radio.'

'Oh sure,' I said, wondering why I hadn't realized that they too use hand held radios.

Once that balloon was erected and the basket upright, Jerry, the pilot, gave the women their final instructions as to where he thought the flight was headed, and climbed into the basket, two way radio in hand.

A short time later, without fanfare, the women let go of the mooring lines and the balloon silently began to rise. At first, the pilot gave the balloon several short blasts of hot air, but as it gained altitude, it began drifting northwest with the wind, towards the freeway. The pilot then turned the gas burner on for a long moment. From where I stood watching, the sound of its hissing gas was like that of a large dragon.

After several minutes the balloon was up about four hundred feet and still climbing. The women suddenly picked up all the gear, throwing it in the back of the pickup truck and got into the cab.

As they were about to leave, I asked the driver where the nearest airport coffee shop was. Her answer was a hearty laugh.

'There's no coffee shop around here. You have to go two miles down that road over there to get to a restaurant.'

I looked at her in disbelief.

'Bye, honey!' she called over her shoulder, as the truck started to move. 'Thanks for helping.'

A moment later the truck drove off the field and onto the access road and was gone.

It was now 8:30 a.m. I walked towards the nearest building, anxious to ask someone where I could get some breakfast. Just thinking about food made me hungrier. As I walked my body felt a little bit warmer. The sun was beginning to take away the morning chill, but even so, I had a feeling that the day was going to be very long indeed.

●

I walked around several large hangars, trying to find someone I could ask about breakfast, but the hour was still too early. None of the hangar offices were open yet. It would probably be 9 a.m. before anyone would come to work. I approached the last building on the field, the FBO that occupied the Terminal Building. The sign on the front of the building read: 'Executive Terminal.' At least I was at the right place, and the building was open. A large hangar abutted the building and was also open and alive with activity.

I walked to the hangar and as I did, I saw two Jet Ranger TV News helicopters parked with their engine cowlings removed, and a King Air. The maintenance operations here were really big time.

Around front, the offices were open in the terminal building, and I walked into the lobby and reception area. A young girl who seemed to have just graduated from high school sat behind the reception counter manning the switchboard. I explained to her that I was meeting a friend here at noon, who'd be flying in. And until then, was there someplace close where I could get breakfast?

'There's no coffee shop around here for breakfast, but upstairs there's a pilot's lounge with coffee and donuts,' she said, smiling.

I thanked her and climbed the stairs, realizing that this field was used mostly for maintenance of corporate aircraft. Several King Airs and a Mitsubishi MU-2 were parked on the front ramp.

The FBO was large and employed a dozen people in its maintenance department, with another dozen in its sales and marketing offices. I was very impressed with the size of its operation, as I climbed the stairs to its pilot's lounge.

The lounge was very nice as lounges go. I expected the usual food vending machines of candy and cookies, but besides those, a large silver urn of coffee and several silver platters of donuts were available to arriving pilots.

After several cups of coffee and three donuts, I began to relax. There were several large picture windows facing the runway along one wall, while overstuffed couches and side chairs and several cocktail tables lined the opposite wall. Large bowls of fruit sat next to the donuts. This upscale lounge was more like what I'd expected.

Talking to several transient pilots that morning, I discovered that this field was one of four that were located within ten miles from each other and ringed Austin in a semi-circle to its north. There didn't seem to be anything very special about any one of the four airfields. Any one was like the other three. The only thing that made this field stand out from the others was that it was closest to Austin.

I'd been at the FBO for about an hour when I decided to walk around the place just to kill time. I went downstairs to go out the main entrance of the building to go to the flight ramp, and everyone I passed in the hallway nodded and asked if I was the guy from California who was waiting to be picked up by a friend flying in. The word really got around fast.

It took me almost an hour to meander around the grounds; I looked over in detail each and every airplane tied down along the ramp. The weather was gorgeous, with bright sun and a few fleecy clouds. The wind would alternately come up and then die down. It was strange, but somehow I still had it in the back of my mind that Les was going to fly in right on time. But Noon came and went with no Les. And now, slowly, the disappointment of not meeting him began to depress me.

There was no place to go for lunch, so I ended up going back to the lounge and getting a couple of apples, more donuts, and a milk from one of the vending machines. It was only then that I actually entertained the possibility that the terrible weather might still cover the entire state, except where I was, near Austin. I was finally understanding that crazy weather like this very well might be the norm around here, while it could be really bad elsewhere. Slowly, I was forced to conclude that it would be foolish of me to expect Les to be able to fly his little airplane through typhoon strength storms and finally land here on schedule. But at this very moment it was really difficult for me to imagine that Les might be struggling through torrential rain, while the sun was shining so brightly here in clear skies above Austin.

I never knew about Texas weather before; nobody ever told me how changeable it can be. I've heard about Alaskan weather, where residents say, 'if ya don't like the weather now, just wait ten minutes and it'll change.' I never believed that there was weather like this in the lower forty eight. I guess I wasn't as well traveled as I thought.

Not knowing about the local weather made me feel like a pilgrim, and I got the feeling from the looks I was getting from the people at the FBO that they thought so, too.

I was getting more and more depressed as I waited for an airplane to land that might never come. This trip probably was a gigantic mistake and I suddenly wished I hadn't come. There was nothing for me to do now, but to kill time until the end of the day. So I alternated between sitting in the lounge until I got bored, to walking around the field, to standing at the entrance to the Terminal Building, and staring outside, waiting for Les.

•

By 3 p.m. something did happened; it rained. The wind came up and then clouds suddenly formed, and within a half hour, the skies were overcast with an eight hundred foot ceiling. Once the sun was obscured, the clouds turned black. Then the temperature dropped about ten degrees, and shock of all shocks, rain began falling. Not a gentle rain, but the kind that Les had been telling me about; the kind he'd been flying through. The rain came down in torrents. It must've rained three inches in less than an hour.

When the rain began I was looking at one of the airplanes parked on the ramp a hundred yards from the Terminal. And even though I ran as fast as I could towards the building, by the time I got there and out of the downpour, I was totally drenched and my flimsy wind breaker was now just a wet rag, hanging limply against my body and dripping water in a pool where I stood.

It was bad enough to be wet, but to be wet and cold gave me a chill and an uncontrollable shiver. Even several hours later, when most of my clothes were dry, my body still shivered.

By 5 p.m. the storm had blown over and the sun came out again. I was chilled to the bone and couldn't seem to get warm, no matter what I tried. To take my mind off my shivering, I wandered outside around some parked airplanes again. The sun was low in the sky as I heard a truck drive up and stop in front of the FBO. From where I was on the flight ramp, I recognized the truck as the same one from this morning; the one with the balloonists. I was too far away to recognize the driver who got out and went into the FBO.

By the time I meandered back to the building, the truck was gone, but as I walked into the lobby I met the FBO's manager, whose name was Bob, who I'd met earlier in the day. I asked him about the truck.

'The Sonnofabitch,' he started. 'I got calls all mornin' from the Tower Chief at Austin, and then from the FAA. The guy that was just here in the truck was the balloon pilot that launched from our field early this mornin' without permission. It seems that he came back to apologize for causing any problems.'

As he talked, I listened and nodded.

'We're so close to Austin's ARSA. He drifted right through it without talking to anybody. No transponder. No radio. He totally screwed up the airline traffic for about an hour.'

'How come he came back here?'

'Well I guess after the FAA got a hold of him, he figured he'd better come over and make up. After all, we got blamed for letting him launch.'

'Well what'll keep him from launching from here again?'

'Not much, except that I told him, if he ever used this field again, not only would we report him to the FAA and local police, but we'd also shoot his ass out of the sky.'

I laughed. I was almost ready to tell him that I was an unwitting co-conspirator in launching the balloon that day, but seeing the look in his eye, I reconsidered and just shook my head, sympathetic to his situation.

•

For the next hour I just dawdled near the entrance of the Terminal, not knowing what else to do, secretly hoping that Les would suddenly fly in and take me away. But, as the sun went down and darkness engulfed the landscape, Les did not come. It was obvious that he couldn't make it, and I knew he wasn't foolish enough to try to fly at night over unfamiliar territory. So I assumed that where he was at the moment was where he'd spend the night.

And one by one, the FBO staff began to leave for the day. First the mechanics left, and then the sales people. And by 6:30 p.m., only Bob the manager and one employee remained.

'Ya know Bill,' Bob said, calling me by name, since we'd been on a first name basis for most of the afternoon. 'Jim here is about to leave to go home, and has to go right by the only motel near here. He could drop ya.'

For a second I thought about it, and realized that if I didn't take his offer, I'd find myself all alone, standing outside locked buildings in the dark, ten miles from town, without a pay phone to call a cab.

'Sure,' I said, nodding. 'I really appreciate the offer.'

And with that, Jim grabbed his jacket and said good night to Bob, motioning for me to follow. It was dark outside and a cold wind had come up. The temperature seemed to be getting down to the forties with the wind-chill. Although I was mostly dry, as soon as I felt the cold wind, I began to shiver. I walked outside, and looked around. The runway lights were on, but nobody was home, and if someone landed in the dark, they'd be on their own until morning.

We walked to the only vehicle still parked in front of the FBO. In my delicate condition of hypothermia, I couldn't believe my eyes. Jim's car turned out to be an open air jeep —

with no top or sides. And where the top should be was a great big thick chrome roll bar. And the seats were upholstered, like they were from a 'Macho Man' car–plain canvas. Not even military jeeps used canvas. The problem with the canvas covered seats was that they were still wet from the afternoon's rain. Needless to say, if I thought I'd been chilled before, it was nothing like the twenty minute ride to the motel at sixty miles per hour in the chilly night air. By the time I got out of the jeep, my body was totally numb.

Jim stopped the jeep in front of the motel and let me out. I tried to give him five bucks for his trouble, but he wouldn't accept anything. I guess all of the locals felt sorry for me. I thanked him and watched, as he drove off into the darkness and out of sight.

•

I took a room for the night at the motel and as I entered its office, I noticed that there was a restaurant right next door. As hungry as I was, I was too chilled to even eat. And I knew that before I could do anything, I had to get warm. So when I got to my room, I filled up the bath tub with hot water and jumped in. I laid there soaking up the heat for a long time and didn't get out until my body was warm enough for my shivering to stop. It took an hour before I felt warm enough to get out of the tub and dress, so I could go next door for dinner.

Dinner was good. I had a New York steak with three bottles of Lone Star. But even after I finished the meal, it didn't seem to help my disposition. Maybe a bottle of whiskey would've helped, but I didn't have one handy.

After I finished dinner and got back to my room, I phoned my wife, Barbara, to tell her about my day. She thought it was pretty funny, as I recounted everything that'd happened to me, since landing in Austin.

'If you think you had fun today,' she chuckled. 'Just wait till you call Les and he tells you about what happened to him, while you were waiting.'

It turned out that Les had called Barbara when he landed for the day, per our agreement. And gave her a phone number where I could reach him. I told Barbara that I thought my entire trip was a big mistake, and that I'd let her know what my plans were the following morning. And after I talked to Barbara I gave Les a call.

He answered on the first ring, and when he heard my voice, he shouted into the phone, 'I'm only a hundred miles from Austin! I flew through some rain today that you wouldn't believe.'

'Oh yes I would,' I replied. 'I was standing outside in it most of it.'

It turned out that he'd been as close as fifty miles from the Executive Airport, but had to detour to a field one hundred miles to the north because of the weather. Flight Service told him that it would be foolhardy to try to make it to Austin with the winds as high as they were. So he detoured at just about the time the rains must have hit the field where I was standing. It was probably the front of the same storm.

But now, he was finally within a half hour's flight of Austin, and if the weather held tomorrow, he really thought he could make it there by noon. I was beginning to understand how 'iffy' the Texas weather can be. And so, we decided that if the weather looked like it was going to clear, then I'd return to the airfield and wait for him. But, if the weather didn't look so good, then I'd just go back to Los Angeles on the next plane out.

And thus ended my first day in Austin.

•

With high hopes for good weather I arose early the next morning and called Flight Service for a weather briefing. I had a feeling that it would be good VFR weather day, but to my astonishment, the briefer told me over the phone that the Austin weather outside was 'zero-zero' in fog. And that the fog wouldn't clear until later that afternoon–and nobody was offering a guess as to what time that would be.

Why did I continue to think that the weather was going to improve? I hung up the phone and went to the window and opened the blackout curtain. Sure as God, there it was, zero-zero. I couldn't believe my eyes. I couldn't see the restaurant next door, sixty feet away. It was going to be another awful day of changeable weather.

Only a fool would hope that Les could fly his little airplane through this stuff. The fog looked bad enough to even ground the airliners; after all, they needed at least a half mile visibility to takeoff, and since I couldn't see across the street in front of the window, I knew nothing would be flying until sometime that afternoon.

I decided to go to breakfast and formulate a plan. But, even though I dawdled at Breakfast, the meal didn't seem to help my logic or my attitude about the situation. The weather situation was rather depressing, so I went back to the motel and packed up, checked out, and called a cab to take me to the Austin Airport. There was little to be done to help things by returning to the Executive Airport and repeat yesterday's scenario.

When I got to the Austin Airport, the traffic around the terminal building was all jammed up in the fog and there were people everywhere. Inside, there were lines of people at all of the ticket counters that were hundreds of people long. There were probably eight thousand people all stranded at the terminal.

At each gate, people were sleeping on the floor next to their luggage, or sitting or standing around waiting, and all as depressed as I was. Apparently, they'd been there all night.

And outside, there wasn't a sound of an airplane engine. I could see through the terminal windows that nothing was moving. The airport was closed until the weather improved.

For once I was glad that I'd eaten breakfast already; I discovered that all of the coffee shops and cocktail bars in the terminal were closed, having run out of food and drink by 8 p.m. the night before. The fog had kept everything from moving. The cars in the streets outside were jammed in gridlock, with no where to go. There was no hope that the food trucks from vendors would be able to get through the sea of cars to restock the terminal building until long after the airplanes were flying again. Until then, nobody stranded here was going to get anything to eat.

The only thing there was to do was wait for the fog to clear.

I stood there looking around and wondered if I should go back to the Executive Airport and wait for Les. But, what was the point? He was stranded on the ground probably by the same fog that now paralyzed Austin. I'd waited yesterday for his airplane, and I knew if I was standing on the ramp all day today, he might not arrive, either. So after a moment I dismissed the thought as a bad idea.

I stood in line and booked a seat on the first airline that had a mid-afternoon flight out of Austin for Los Angeles. For once Flight Service had been right about the weather. And, if they were right about the fog clearing by afternoon, I'd be glad to get out of Texas. And I felt pretty confident that in several more hours the RVR would be at least to a half mile, enough for the 'smokers' to get airborne.

As I waited with the thousands of other travelers I wondered if it had all been worth it–not only the expense of commercial airline tickets both ways, but a motel for a night and two meals.

Suddenly, I really felt like a 'Pilgrim' on his 'first cruise.'

Day Eleven and Twelve

I went back to Los Angeles the afternoon of day eleven of Les' journey back to California. He told me later, that it took him until almost 6 p.m. that day just to fly the one hundred miles to the Executive Airport, so that if I'd waited there at the field until dark and then left, I still would have missed him. He said that he couldn't takeoff for Executive until late in the afternoon, when the fog finally cleared. We must've both taken off at the same time that day, Les for Austin, and me for Los Angeles. The one piece of good luck Les had that day was that the weather stayed VFR well into the evening, so he was able to fly on to Midland before quitting for the night. I guess he just landed and took off again, when he didn't see me waiting.

The next day, day twelve, was the day when he finally got out of the storm track after leaving Midland. Day twelve was a very long day for him as it was the last day, Sunday, of his two week vacation. He had to be back to work in Los Angeles the next morning. After days of flying just a few hundred miles or waiting on the ground for storms to blow over, he was determined to fly all the way home and park his airplane that night in California.

The day started out with him taking off before first light from Midland, and flying all the way to Hawthorne, California, by 10 p.m., with fuel stops at El Paso, Tucson, Blythe, and Palm Springs. It was a fourteen hour flight because of strong head winds. His ground speed coming from El Paso was down to eighty miles per hour.

•

In a way, my being unable to ride back with him was a blessing in disguise. After flying in his Cadet months later on several short hops of an hour or so, I discovered that there was less leg room in his cockpit than in my Pacer. And my legs were totally cramped after just a few minutes of sitting in the Cadet. It turned out that the Cadet has the same size cockpit as the Ercoupe.

Chapter Five
In The Soup and On The Gagues

Flying Blind With No Peeking

For the past twenty years I've been flying strictly by Visual Flight Rules (VFR). That means that when the horizontal visibility goes below three statue miles, and/or the ceiling drops below 1,000 feet, sometimes referred to as 'going below minimums,' I can't fly and spend the afternoon sitting on the ground waiting for conditions to improve.

In the Los Angeles basin the weather usually goes below VFR minimums when the cloud deck is only a couple of thousand feet thick. The smog layer normally climbs only to eighteen hundred. So after years of being constrained by Instrument Meteorological Conditions or IMC (where conditions are less than VFR), I thought it was time to get an instrument rating. With the rating, I could go down to the airport and fly any time I pleased. Of course, every Pilot's dream is to be instrument rated, but getting the rating isn't easy or cheap. It takes the sacrifice of time and money and for a family man may be a 'bridge too far.'

When my involvement with instrument training was over, I learned that the training takes at least three to four flights each week with an instructor just to develop one's instrument scan; and then several very large wheelbarrows full of money to pay for those training flights (at about sixty dollars per hour for the airplane (if you don't own your own), plus twenty five dollars per hour for the instructor. That's almost four hundred each week for about three months. That total approaches the number five thousand real quick.

The real hang-up with the rating for pilot's with families is that when the flyer finally has the time, he never has enough money. So as long as I had to rent airplanes I could never get ahead of the game. And it wasn't until after I owned my second airplane, which was correctly instrumented and certified for IFR flight that I finally had the right equipment and enough money at the same time to afford the training. But, real life being what it is, having the equipment and the money wasn't the whole equation; having the time, was quite another matter.

From April of one year through October of the next I trained seriously for the Instrument Rating. And as I fantasized about it in the beginning, with the rating I could go down to the airport and takeoff in my airplane regardless of the weather, short of thunderstorms or hurricanes.

For VFR pilots in Los Angeles, the problem is getting out of the basin. Instrument pilots, of course, just ask for an instrument climb to 'VFR On Top.' And, once out of the basin, the remainder of California is normally VFR.

When a pilot just flies for pleasure from airport to airport, sitting on the ground waiting for better weather is taken for granted. But, waiting can have a serious impact on a VFR pilot, if he is trying to depart on a trip with his wife. As a matter of fact, one of the few reasons wives even consider traveling with their husband-pilot in small airplanes is simply to avoid the hassle of flying commercially, and then, only for short trips of about three hundred miles.

Most wives don't like to fly in small airplanes. After years of pondering this, I concluded that it was because non-pilot wives are usually the pilot's 'first wife.' And, there seems to be a universal law that says, when a girl marries a general aviation pilot as his 'first wife,' it's imperative that she be afraid of flying in small airplanes. That truth holds in almost every case I've studied. Now, in cases where a non-pilot wife loves to go flying with her husband it's always a second marriage (where the non-flying spouse understands that flying is a large part of the other's life, and is to be mutually enjoyed, or no marriage).

Another reason women don't like small airplanes is because they don't get food , alcohol or a bathroom to use while flying. Airliners are the only vehicle that affords such luxury. The trouble is that flying on the airlines these days is not only an inconvenience, but is also a waste of time and money.

The little woman wants only to get underway immediately. But, if flying in a small airplane means that she has to wait at some neighborhood airport for the weather to improve, and there isn't a nearby bar or restaurant to while away the time, then she might as well put up with the commercial hassle. At least at the commercial airports she can sit and readjust her attitude with a glass of wine.

But the dream of traveling by small airplane when the weather gets 'iffy' can be realized only if the pilot is instrument rated. Without the rating, the pilot must plan to depart at an hour of the day when he is certain that the fog will be gone. And at most small airports, that means around mid afternoon during the wintertime in California. For the VFR pilot, during the wintertime, daylight is at a premium, and if he can't takeoff until after lunch, then he really has only a three hour window of flying time before dark. And if the flight is longer than three hours, women aren't sure they want to fly in small airplanes in the darkness of night. And, a smart VFR pilot should consider any night flight to be an instrument flight.

Trying to launch by a specific time always puts a lot of stress on a pilot, no matter the reason. I know, because it sure stressed me the one time my wife and I launched on a trip to the fly-in resort of Punta Chivato, Mexico. What happened is described in Chapter Seven.

•

Besides the weather affecting the VFR pilot, flying at night in total darkness (with no moon) can disorient the most experienced pilot. Flying on a moonless night means that there will be no perceivable horizon for the pilot to keep his airplane right side up. Without a horizon a flyer must trust his flight attitude entirely to his instruments. And to develop an adequate instrument scan enabling a pilot to correctly interpret those instruments can only be developed over time with expert training.

To illustrate why flying under visual flight rules on a moonless night in total darkness is risky business, and is really an instrument flight, I offer the following night flight my wife Barbara and I made one summer evening.

The town of Solvang is nestled in the middle of the Santa Ynez Valley, just on the north side of the coastal mountains inland from Santa Barbara. We'd flown up from Los Angeles in our Pacer on Saturday morning, and after a wonderful weekend, we were now ready for the flight home from the airport located south of the town of Santa Ynez, near Solvang.

The flight back would take an hour and forty minutes, and as I looked up at the setting sun, I knew that darkness would overtake us at the halfway mark, somewhere around Point Mugu. So, even though the light was fading, before taking off, I elected to top off the fuel tanks.

Of course, Barbara wanted to get going immediately, but I reminded her that by the time we'd flown halfway home daylight would be gone, and for our safety, I wanted to have full fuel.

When I told her we'd be flying in darkness she began to get nervous, and since I knew that she didn't like to fly period, and only went along this time so we could visit long time friends from Northern California that were vacationing in Solvang. To put her at ease, I gave her the option of spending the night in a motel in Santa Ynez and flying home in the morning. The decision was hers. And, after a moment, she decided that we would fly back that night.

And, so I 'topped' the tanks, taxied out to the active, and took off on Runway 26, just as twilight was beginning to fade. As we cleared the San Marcus Pass, heading for the coastline of Ventura, thirty miles in front of us, the shades of night were growing longer. Although the flight was uneventful, as we approached Point Mugu, as if on cue, darkness fell.

•

Usually, flying at night in the Los Angeles basin is simple. With fifty square miles of lights in all directions, you really have to work at getting lost. Just keeping 'the blue side up' is also no great trick. So what I expected to see on the flight home was millions of lights in the distance, lighting up the night sky with a bright glow of 'city shine,' that could be seen all the way to Santa Barbara. I expected to home on the distant glow, making navigating easy.

But what I expected, and what I got were two different

things. Looking eastward from Point Mugu to Santa Monica seventy miles away, I could see nothing but blackness. No lights in the distance, and no moon above. Just darkness. There seemed to be a shallow layer of haze in the air that was obscuring the few lights that should have been visible. When darkness came, it quickly descended like a curtain; and then everything faded to black.

Traveling the area from Point Mugu to Malibu in a Piper Pacer at a speed of 103 mile per hour takes fifteen minutes. The terrain has no recognizable landmarks since it's a wilderness of coastal hills until Point Dume where Pepperdine University is located. Along the route, the coastal hills rise up from the water's edge to twenty five hundred feet, on the inland side of the Pacific Coast Highway, as it meanders along the rugged coastline.

The only visible lights were from off the shore to our right, where a line of oil drilling platforms stood in a row, each exactly one nautical mile apart stretching from Point Mugu seaward towards the Channel Islands. There were fourteen platforms in all, although only the lights from several of the closest ones could barely be seen. And as we flew by, the few pinpoints of light from the platforms rapidly faded from view. On a night like this, there wasn't much to see from an altitude of three thousand feet. It was like being a fly, buzzing around a dark closet with no idea of up from down.

For Barbara, the fact that total darkness offered us no idea of our attitude was frightening. If there'd been a bright moon, bathing the waves with glistening highlights, the flight would've seemed peaceful, relaxing or even romantic — like all of the night flights I'd flown before. But the flight tonight was different, complete darkness, with no moon or stars and no way of visually telling up from down. Our mood was somber and we flew in silence — the kind of silence that, if broken, might lead to something terrible. So, one tries to remain quiet. But, suddenly I felt Barbara shifting in her seat next to me. She was restless, and I took that as a sign that she doubted the successful outcome of our flight.

The total darkness could have made me fearful as well, but I took solace in my flight instruments, which unemotionally and continuously indicated the airplane's attitude. So, in an attempt to ease Barbara's mind, I began explaining to her, over the intercom, the function of each instrument in the panel before us, and what information each was giving me about our situation. It was sort of a one minute tutorial that a pilot might give to a small child during his first airplane ride. And, I tried very hard to phrase the explanation in a way that didn't sound too simplistic or patronizing. I summed up by trying to define attitude flying in just a few sentences, hoping it would alleviate her concern. But, I'm not sure it made her feel any more comfortable.

A pregnant silence followed my explanation, and we each occupied ourselves with our separate thoughts, as the Pacer droned faithfully onward through the night. After a moment, I looked up through the top of the windshield, searching for stars, but there was only darkness. The sky overhead was obscured by a high overcast making tonight's flight a dark and lonely trek.

As we cruised through the night skies there was no sensation of movement. We sat belted into our seats as if we were at home sitting in our living room. The only sensation was the engine's noise and vibration, which was greatly attenuated by our hard headsets covering our ears. We sat in our seats as if suspended in our little 'island in the sky.'

As we rode along, it was easy for me to understand, how in 1935, more reflective aviators like Ernie Gann could relish this kind of peace while flying through the night in a DC-2, from Newark to Cleveland, on American Airlines route AM-21. And, with nothing much to do in the way of flying chores except to monitor the airplane's instruments, a pilot's mind is freed up to reflect on life and its vicissitudes.

No matter what size airplane a pilot flies, the experience can be either enlightening or sobering, when considering one's being in relation to God or the Cosmos. In some ways, I'm glad that, for me, the experience of flying, itself, is enough to ponder, since I've already understood and accepted my mortality.

•

I never realized before tonight's flight how disorienting complete darkness can be. And for a VFR pilot, if the condition of total visual blackness were allowed to continue for more than a few seconds, disorientation is immediate. Within seconds, without instruments for reference, a pilot would have to depend on the sensations of his body to indicate which way 'up' is. But, the body lies; it relies on the sensation that forces put on it, like positive and negative 'g's.' And, when the wings are not kept level a slight turn develops, and, if left uncorrected, seconds later the turn evolves into a 'graveyard spiral' If that spiral is left uncorrected, it always ends in a scene of 'balled aluminum, scattered over all quadrants.'

Tonight, the instant I realized there was no horizon I was honest with myself and acknowledged that we were flying in instrument conditions, and that although I wasn't rated, I'd better begin the best instrument scan I could muster. So I focused my attention mostly on the attitude indicator, which became my 'small windshield' to the world, since my larger windshield had no view.

There was no wind, so trimming the airplane for level flight was an easy matter, and once trimmed, left little for me to do except to continue my instrument scan. Because of the short distance between Point Mugu and Malibu, I was only on the 'gauges' for fifteen minutes.

The long minutes passed with no outside view. Suddenly, clusters of lights from the direction of Santa Monica began appearing through the mist. I was quick to point this out to Barbara. And, as the number of lights began multiplying, Barbara slowly began to relax and stopped squirming.

I glanced directly below us through my side window, thinking that I should be able to see some light around Pepperdine. Finally, after several quick glances, I was able to make out the bell tower of its chapel. As soon as I knew we were over Pepperdine, I relaxed and gave up my instrument scan, and looked over to Barbara. 'We're over Pepperdine, and we'll be on the ground soon,' I smiled.

She nodded, staring straight ahead, still not convinced that the danger was past. She squirmed several times, but as more and more lights became visible and she was able to tell 'up' from 'down,' then she looked over at me and smiled.

Ten minutes later, we were abeam Santa Monica, along the coast, and as we turned southeasterly with the shoreline, and headed towards Marina Del Rey, it became evident why there was no 'city shine.' The entire western half of the basin was overcast. A low deck of clouds, now beginning to scatter, lay in a shallow sheet at two thousand feet.

It was at that moment that I decided to start working on my instrument rating. I thought that it would be the only way I'd ever gain Barbara's confidence about flying in small airplanes, especially ours. I hoped that in retirement we could use our airplane like a recreational vehicle and do some long distance travel.

•

Since Torrance Airport is intermittently below VFR minimums during winter mornings, I thought I'd start working on my instrument scan there. I wanted to begin at a leisurely pace. So, I planned a one flight per week frequency. I reasoned that once a week would get my airplane and me exercised regularly, and that after a while, when I'd developed a pretty good scan, then I'd decide whether I wanted to go all the way for the rating.

Looking back, I was totally naive about what it takes to get the rating, as far as the amount of time and money. It wasn't until a dozen months later and one hundred and six hours of 'instrument time' in my log book, that I realized the truth; but by then, I was out of time, money and ideas, and had to quit. And it was only after I'd spent just under six thousand dollars and exhausted my funds that my CFII instructors finally leveled with me as to what the rating required in the way of training commitment.

To get a good scan perfected to the point that a pilot could fly 'hands off', and open maps and change radio frequencies without the airplane moving more than one or two degrees off course, turned out to be a function of the frequency of lessons, not the total training time. Apparently, an instrument scan improves only if the student trains often, at least three times each week, until he's mastered the curriculum. Actually, training everyday until the checkride is the optimum.

As I began my instrument training, I also got really busy at work, and by each weekend I was too tired to train, and should've been smart enough to throw in the towel and quit right then, before I really got started. I could always train another time when I wasn't so busy and was rested. But the instructors, sensing that their 'cash cow' was leaving, kept encouraging me to continue.

•

Guy Gillmore

His name was Guy Gillmore and he was sixty six years old. He was the only CFI on staff at Southwest Skyways, the oldest FBO at Torrance Airport. I had taken my checkride for the Private License there some twenty years before, and since then, from the looks of things, nothing had changed. The old rusted corrugated metal buildings had been repainted, but the years had not been kind to them, and they looked their age. Over the years, I'd rented airplanes from the FBO, and that's why I was there now; living only a few miles away made it convenient, so I stopped in to talk to someone about the instrument rating.

The first time I saw Guy, he was sitting on an old couch in the pilot's lounge. One look and I almost laughed; he was a look-a-like to the then President Ronald Reagan, right down to his dark horn rimmed glasses, dark hair and deep voice. He wasn't as tall, but that was the only difference; his voice was identical and friendly as the President's; and as a final touch, his face was the face that had weathered sixty plus years of sun and was just as 'leathery' as Reagan's.

After we shook hands, I got right to the point, telling him about my worries as a VFR pilot. I told him that I'd like to fly just once a week, on Sunday mornings; that I wanted to ease into instrument work, and not get too serious about it at the start. First, I wanted to develop a scan that would save me in an emergency, and, after that, when I got a good scan going, then I'd decide whether to pursue the rating.

That all sounded fine to Guy, but, he cautioned me, to develop a good scan would take about thirty hours. He thought that by then, we would've covered a lot of the FAA instrument curriculum, anyway. He thought that the time spent wouldn't be wasted.

So beginning on the very next Sunday morning at nine a.m. Guy and I began my training. And for the next fifteen Sundays, we would launch on an IFR training flight that always began by flying from Torrance Airport to the training area off the shores of San Pedro. The training area was just south of an electronic navigation intersection named 'LIMBO.' And sometimes when my lessons weren't going especially well, it was hard for me to decide whether I'd been at LIMBO, or was just 'in LIMBO.'

Geographically, LIMBO was located two miles off the coast of Palos Verdes Peninsula, just off a point of land that once was the home of Marineland, — a large aquarium complex and a popular tourist attraction during the seventies and early eighties. The intersection is a point in space at an altitude of three thousand feet. There, two electronic beams intersect, one from the Seal Beach VOR station some twenty two air miles to the east, and a beam from the Los Angeles VOR station some thirty air miles to the northwest.

Once we got to LIMBO, then we'd proceed to wherever our airwork took us that day. Sometimes it would be holding patterns over LIMBO. Holding Patterns are race track shaped maneuvers that are used to slow up an airplane's progress during an instrument flight, for greater separation between airplanes. Sometimes, it would be precision ILS approaches, or non-precision approaches using the automatic direction finder.

We started off with straight and level flight under the hood, followed by turns and climbs and descents. After a month or so I developed what was to be my instrument scan. And it wasn't long before I realized that learning instrument flying was a lot like learning Mathematics; learning it was mythology and art, within a framework of rules. Whether a teacher of Mathematics is ever successful at making a student understand concepts is something left pretty much up to the student.

In teaching instrument flying, the teacher has little to do but to point out to the student what instruments he should be watching at any given second. So that the important part of instrument interpretation is left mostly up to the student to figure out. And the instrument scan, it turns out, is a hierarchy of specific instruments to be watched in a specific order. The problem is that the order is maneuver-dependent, and changes with each discrete flight maneuver. So that instruments of primary importance in one maneuver are different from the primary instruments to be watched in another maneuver.

Immediately, I could see that it would take forever to know what instrument to watch for every different maneuver. It seemed almost insurmountable that a human could master the complexities in just thirty hours. What I should've realized, quickly, was that it would be impossible for me to learn all that over time with just one lesson per week.

But since my premise was that I wasn't going to get too serious about the rating, too soon, and wanted only to learn enough for 'life insurance,' the fact that the scan might be really tough to develop didn't bother me. After all, the other half of why I was flying every Sunday was just to get my airplane in the air on a regular basis, for its general health. Otherwise I'd be sitting on the ground waiting for the fog to clear with all the other VFR pilots.

Since I didn't seem to be getting the 'hang of things' as fast as I thought I should, I began reading everything on the subject that I could lay my hands on. I was looking for the 'secret' that would allow me to break through the plateau that held me captive. By the time I finished a dozen books and manuals, some old, some brand new–they all said the same thing: there's no royal road to developing an instrument scan. It's up to the individual and it only comes with frequent practice.

That sounded like a big cop-out to me. Just like learning mathematics; nobody can really teach it, the student just has to pick up the concepts on his own. I wasn't too impressed. Even the instructors I talked to pretty much all said the same thing. 'It's something you just have to develop on your own–everyone's different–it's such an individual thing.' Even Guy told me on numerous occasions, 'It'll come to you at about the thirty hour mark.' I think he told me that because the thirty hour mark was when it all came together for him. What he didn't realize was that if it came to him at thirty hours, it was because he was training everyday. If he'd thought about it in view of my training regime, the number was probably three hundred hours.

So every Sunday without fail, Guy and I would takeoff and end up at six thousand feet over Catalina Island, or over the channel, halfway between Catalina and Long Beach. And everything we did in the air was from miles offshore. Sometimes the channel below was clear, but mostly it was a thick deck of clouds that topped out at about five thousand. Even the time that Guy was teaching me to use my ADF to track to a station, we were over Avalon trying to track the signal of KNX Radio; it's antenna is located fifty miles away in Torrance. No matter what the maneuver, we did it from the Island.

Finally, at a certain point in my training with Guy, when I was really tired, and I'd show up and he'd ask me what I wanted to work on that day, and I'd just say whatever I felt like, and we'd do it. And when I didn't feel like doing anything at all, I'd just say, 'holding patterns.'

LIMBO is a garden variety intersection with a published holding pattern. My problem with it was that the winds there were so strong that I usually had to add or subtract up to two dozen degrees of correction angle to the published headings. And in my exhausted state I could never remember what the original published inbound or outbound headings were, let alone, the additional correction angles. The bottom line was that I really didn't learn very much about holding from Guy or LIMBO.

It's too bad that I didn't get to LIMBO on days when there wasn't any wind, as I'm sure I might've understood, at last, what it felt like to fly a published hold and execute it correctly. But by the time Guy and I 'called it a day,' I think I must've logged seven hours of holding at LIMBO, which equated to four hundred and eighty three dollars of flight operating costs at the rate of one dollar and fifteen cents per minute.

•

By my thirty three hour mark I'd reached a point in my 'scan' beyond which I didn't seem able to improve. And although Guy kept telling me that I'd get better with practice, I remembered his magic number of thirty hours and my scan wasn't as wonderful as promised.

At that point I was tired of being Guy's 'cash cow.' And I also realized that if I suddenly did want to develop a good instrument scan that I'd probably have to do it with another instructor. So, one day in May, I phoned Guy and told him that I decided that I needed to improve my scan before I wasted anymore of his time. And, I thought I'd try to perfect it with some simulator work. It wasn't a lie, really, but it was the only painless way I knew of for us to 'split the blanket.'

•

Per Oman

One afternoon, several weeks later, while driving home

from work, I found myself on the perimeter road of the Torrance Airport. As I passed several FBOs located about a half mile east of Southwest Skyways, an FBO called Benbow Aviation had erected a new banner over its doorway, proclaiming 'Frasca Flight Simulator.' I didn't travel the perimeter road much and so I hadn't seen the sign before, but hoping that a couple of hours of simulator time might smooth out and improve my scan, I stopped to inquire about the Frasca.

Per Oman was the name of the instrument instructor at Benbow Aviation. He was six feet three inches tall, and weighed about one hundred and seventy pounds. He appeared to be on the thin side for his height like a basketball player. And beside his thick accent, his bushy blond hair and blue eyes verified his Swedish heritage. I never knew his age, but he appeared to be in his early thirties.

I shook his hand and explained my problem. That after thirty three hours with Guy, not only was I disappointed with my progress, but I was out twenty three hundred dollars, and all I had to show for it was two pages of entries in my log book.

Per walked out from behind the counter and took me outside. It was late May and the sun was just disappearing below the hills of Palos Verdes, to the south. As we stood in the fading light he told me that he understood perfectly what I was talking about. It seems that a year ago, he and Guy had been the only CFIIs at Southwest Skyways, and that the reason he left Southwest was because he didn't agree with their policy concerning the training of instrument students.

When I heard that, I made an immediate appointment to fly the simulator with Per the next evening, so that he could evaluate my skills for attitude flying.

•

I arrived at Benbow Aviation the next evening at seven p.m. When I got there, Per was all alone; all of the day students and staff had gone home an hour before. Per was sitting in the back office waiting for me. He ushered me into the simulator room.

The simulator looked very realistic, and was built as if it was the front half of a flight deck of an airliner. It had an enclosed fuselage from the windshield and instrument panel back about four feet, enough space for a chair to be placed in front of the instrument panel for the pilot.

Behind the frame of the windshield was a projection screen, so that during the takeoff and landing phases of flight a computer-generated picture of an airport's runway appeared to make the sensation more realistic. During the enroute portion of flight, the picture of an undercast of clouds filled the view through the windshield.

The effect was complete with sound effects for the engine and tires screeching during the touchdown. The only thing missing was motion.

Off to one side, and behind the pilot's seat, was the instructor's station, comprised of several large panels of switches and a large X-Y Plotter to trace and record the progress of the flight with an ink pen on an enlarged enroute instrument chart of the Los Angeles area.

As I studied the simulator's instrument panel, I immediately realized that I was going to have big trouble trying to fly instruments that were more sophisticated than I'd ever seen before. The simulator was actually portraying a twin engine cockpit; it had two of every lever and knob. The instrumentation was for high performance twins and had multiple use indicators like the Horizontal Situation Indicator (or HSI) that combined four instruments into one. Very few general aviation pilots flying low performance, single engine airplanes would ever see an HSI in real life and I was living proof.

Never having seen an HSI close up, I had no idea of how it worked or how to interpret its indications. The more I studied the simulator's panel, the more I began to feel that this flight wasn't going to be any kind of example of my real instrument skills, such as they were.

I mentioned my concern about exotic instrumentation to Per, but he dismissed it with a laugh, telling me not to worry about it. He said that he'd reconfigure it as a single engine airplane. Nevertheless, those extra knobs and levers were in my way and would only help to confuse me.

As he set up the simulator and reconfigured it from the instructor's switching panel, he began to explain the flight that he wanted me to fly. It was fairly simple and straightforward beginning with an instrument takeoff and climb from Torrance Airport to LIMBO intersection. Then, he'd give me vectors to fly, and I'd do some airwork, like VOR tracking, and I'd conclude the flight with an ILS Approach into Long Beach Airport. And as he finished flipping switches and turning rheostats, he explained that every inch of the flight would be recorded by the pen of the X-Y Plotter, so we could critique everything I did, right and wrong, at the flight's conclusion.

It all sounded so simple, but I could tell that I wouldn't do well, as I sat there staring at all of the extra dials. I kept thinking it's too bad that the panel wasn't as simple as the one in my airplane. But, there wasn't anything I could do about it now. My skills were going to be evaluated at the rate of a dollar per minute, while trying to fly a simulator configured with stuff I'd never seen before. But, since I was already committed, I couldn't back out without looking like a complete incompetent.

When I warned Per not to expect too much from me because of the sophisticated instruments, he just looked at me like he thought I must be a moron, but a moron with money. I guess he thought he might as well humor me.

The climbout went fairly well, once I got the hang of it. All simulators take about two hours to get used to, but two hours was all the time I was going to get for the evaluation. I wasn't sure that a 'Frasca flight' would really provide the good data he'd need to evaluate me, and considered this flight to be my impression of how the Marx Brothers flew instruments. Only, instead of 'A Night At the Opera,' it would be 'A Flight To LIMBO.'

The LIMBO intersection consists of the intersection of two electronic beams coming from two separate VOR stations; one beam from Los Angeles and one from Seal Beach. To navigate to LIMBO, one takes off and climbs out on the runway heading, contacting Los Angeles Departure Control when told to do so by Torrance tower. Departure Control then gives the pilot one or two headings to steer, using radar, called vectors, so that the airplane intercepts the 170 degree radial from LAX VOR at an angle of thirty degrees.

The pilot watches the needle of the VOR indicator set to the 170 radial of LAX, and he continues to fly until the needle slowly centers, as the airplane approaches the radial. When the needle centers, he then turns the airplane to a magnetic heading of 170 degrees and tracks the radial southward. While tracking the LAX radial, the pilot watching the second VOR indicator that's tuned to the 251 degree radial of Seal Beach. When the needle of the second VOR centers, the airplane is at LIMBO.

Of course, that's how it's suppose to work, but as soon as I flew through and past the 170 radial of LAX, I knew I was in trouble. When the needle slowly began to move across the indicator, I didn't react. Instead, I just stared, mesmerized. From that point on, I was playing catch up, chasing needles. If you're flying instruments in an airplane and chasing needles, you'll get into real trouble, real fast.

As soon as I'd busted the LAX radial, Per suggested that I begin a turn to the left to re-intercept it, then track it to the intersection. That was a wonderful idea, and one that occurred to me right after I flew beyond the radial. But knowing what you should do, and then doing it, are two different things. The difficulty came when I tried to track the radial using the HSI.

The HSI indicator simultaneously depicts the airplane's heading, course, and bearing to the VOR station being used for radio navigation; using a series of flags, needles and a minuature airplane in the instrument's center. It really indicates in real time the airplane's horizontal situation regarding its flight over the earth.

The HSI is a fascinating instrument to watch, because of the many moving parts with a life of their own. I fixated on one of its needles, beginning instrument students are warned never to do. The HSI was doing its best to tell me where I was, but I couldn't interpret its message.

About the time I finally recaptured the 170 radial from LAX and was tracking it perfectly, I'd already flown past the 251 radial from Seal Beach. The meant that I'd already passed the intersection, and was headed south of it towards another intersection by the name of 'PADOR.'

Now, as if that wasn't a bad enough problem, I suddenly realized I also had an altitude problem. I remember being cleared to climb and maintain three thousand feet on the way to LIMBO. But since I was so fascinated with the HSI needles to 'nail' the intersection, I hadn't bothered to scan the altimeter. I was now climbing through four thousand eight hundred feet.

It was like that for the rest of the flight. I was a nervous wreck the entire time, trying to catch up with the airplane. And even though I'd logged over thirty hours of time on the gauges with Guy, I was acting like a new instrument student on his first flight.

Finally, Per gave up, and decided to set me up for the ILS approach into Long Beach. I think he did this to see if I could possibly top what I'd already screwed up, so far. I know he expected to see the movements of the Glide Slope needle and the Localizer needle, both contained within the ILS indicator, appear to have a 'sword fight' with each other, as they moved in and out of each other's way, during the final approach from the Outer Marker.

Of course, I didn't disappoint him; he got his 'sword fight,' but I was also able to get the approach all together by the Middle Marker, through a series of 'horsing' maneuvers with the airplane, banking it first this way and then back; like a ship in a storm, rolling first thirty degrees to the right, and then thirty degrees to the left.

Either Per saw some glimmer of hope in my 'deft touch' during the landing or maybe he felt sympathetic at my first confrontation with an HSI or twin engine panel, or maybe he just realized that this 'moron' had more money than brains. Anyway, the bottom line was, I was an additional 'living breathing' instrument student he could add to his list.

After I'd spent almost an hour on the simulator, Per critiqued the flight. He said it was difficult for him to evaluate my skills, because he couldn't tell whether my poor performance was due to my confusion with a sophisticated multi-engine instrument panel, or lack of instrument skills.

When I explained to Per that I was a much better instrument flyer in my own airplane, he agreed that the only true evaluation of my skills could come during a flight in my own airplane. It would be the acid test. And that was fine with me, since flying a real airplane is always easier that flying a super sensitive training simulator.

•

So, the next evening at seven p.m., Per and I found ourselves climbing out of Torrance Airport in my Musketeer, bound once again for LIMBO. Needless to say, I was a lot better in my airplane than I'd been in the simulator. Of course, I was still pretty rough with my scan, something I was not happy about.

The flight lasted an hour, all under the hood, doing airwork, mostly. It ended with an ILS approach into Torrance. And although Per setup the radios and VORs for the approach, my execution of it was adequate, he told me, for a 'ten hour' student, but not for someone with thirty three hours.

After Per's critique, I decided to continue my instrument lessons. His assessment of my skills seemed to be right on, and his plan to bring me up to speed in short order. seemed reasonable. His confident tone made me feel that not only could I learn what I should have from Guy, but Per had his own rigorous format for setting up avionics for an instrument approach; Guy never seemed to have any rigor in his method at all

It turned out that Per was booked on the weekends, so the only time he had available for a new student was during the week. So we started flying in the evenings. And during my second instrument lesson at night I realized just how much easier it was to fly instruments after the sun went down. Flying approaches and holding patterns are easy when there's no wind to worry about. With no wind, the headings are always the same as published.

After a month of flying at night Per's daytime schedule loosened up so that we could fly weekends. We switched to Saturday and Sunday afternoons. And, as the months went by, flying on both days of the weekend soon became laborious and too much of a grind for me, especially after working all week with no days off. Instrument training was not fun; it exerted a lot of pressure on a pilot to concentrate. I was so exhausted that I should've just quit cold right then, but Per, always the optimist and full of energy, dragged me along in his spirit's wake.

By this time, I'd been exposed to the entire curriculum for the rating, and although I was still at a plateau, the plateau was a little higher than the previous one with Guy. My additional practice was paying off.

Per and I had started flying in June, and now it was close to Thanksgiving. And in all the time we'd been flying together, Per never mentioned that his work visa was expiring in early December; then he'd have to return immediately to Sweden. What he didn't mention was that, after three years in this country trying to make a living teaching the instrument rating to pilots, he was starving to death. The economy in California 'went south' and with the failed economy, most of his students couldn't afford to fly.

Per decided that since I'd completed all of the FAA curriculum for the rating, we should do the long instrument cross country flight. Then my next instructor could simply review everything and polish me up for the checkride.

So on a Sunday morning, Per and I took off on the long instrument cross country, from Torrance to San Luis Obispo and back. It was a tiring flight flying instruments for five and a half hours, but when it was over, it felt good to get that behind me. After the flight, Per gave me his usual half hour critique. He ended it by saying that if I could fly more often during the week, then my skill level would really improve, and from there it would only be a couple of weeks of review before the checkride.

It sounded so good, but he was leaving forever, going back to Sweden. So I took his lecture as just motivational encouragement. I knew that I wasn't anywhere near ready for a checkride anytime soon. So, at the end of thirty three hours of hood time with Per, I had to admit that I had learned a lot of things that Guy never taught me.

And, here I was again. It was Christmas and Per was gone, and I'd spent almost fifteen hundred dollars with him, and still no rating. All I had to show for the time spent were entries in my logbook, now totaling sixty six hours of instrument time. Per's training had given me greater confidence as a pilot; and I'm grateful for that. But, still, it wasn't enough — I wanted the rating.

•

Ground School

Four months of brooding and depression went by. I had no plan and wasn't sure what to do. Guy had 'flimm-flammed' me, Per had abandoned me, and now. I needed to keep motivated about the rating. I realized that with all the flying I'd done in that regard, I'd never gotten around to taking the written examination. It is the most difficult test the FAA offers and over the years, it had taken on a life of its own in my mind.

I'd tried on four separate occasions to cram for it, but at the last minute something always came up. I just couldn't do it on my own. I needed help.

I decided that my only hope was a weekend ground school, where for three continuous days, you learn everything you need to know to pass the test. And on the Monday morning following the three days of classes, you're tested.

So I found a ground school of international renown, that was located in Santa Monica, only twenty miles away.

I phoned and was surprised to learn that their next weekend

Ground School for the Instrument Rating was only three weeks away. I signed up immediately.

And, after a whirlwind three day cram session, concluding on Monday morning's computer graded test, I walked out of the school with a certificate of completion. And, I must say, it sure was a good feeling to get the written behind me.

•

One of the little extras that the ground school gave each of the students who completed the course was a free two hour ride in the Frasca 121 Simulator. The school had two 121's located side by side in a room just off of the school's foyer. The simulators were visible to all entering the school's lobby and gave the place a professional air. I was interested in flying the simulator because during coffee breaks over the weekend, I would wander down and look at them and even spent ten minutes sitting in one.

I couldn't believe it–the instrument panel of the simulator was outfitted almost identically to my Musketeer, except that I had several more radios in my airplane than the simulator. The layout of the panel was copied right out of my cockpit — and the thought suddenly occurred to me that this was one simulator I could master.

I made an appointment to take a two hour ride in the Frasca 121 the very next weekend. And, after the ride I began to think about getting back into finishing the rating. If I used their school, I could fly my own airplane with their instructors, and park it on their ramp at no cost. All I had to do was sign up with the school and bring my Musketeer over from Torrance. Since they had plenty of instructors on staff, finding someone to fly with me couldn't be a problem.

So the next weekend found me winging my way to Santa

Monica Airport and parking my airplane on the School's flight ramp. Once I'd done that, I was scheduled to begin my flight training the next weekend.

•

Charley

Charley, my instructor, was from Belgium. 'Charley' wasn't his real name,which no one could pronounce. He was thirty years old and was in the States on a work visa.

Charley was almost six feet tall with a thick head of coal black hair that he wore in a crew cut. His complexion was dark from the Italian and German influence on the French, all mixed together. He was from a town near the border of France, and he said that most of the people living in border towns took on the physical attributes of the neighboring country.

In his white shirt and black tie, the uniform that all of the CFIIs wore at the school, Charley look liked one of Hitler's finest. And though he kept reminding me that he wasn't German and that his countrymen were a cross between Germans, Italians and the French, his thick German accent gave him the perfect image of a retired WW II U-Boat commander. But, he had a good sense of humor and was a terrific pilot, so I thought that this time, I really had a chance to finish the rating.

For the first lesson, I expected to be flying my airplane; but the flight school trains everyone in the simulator for the first twenty hours.

For the next three weekends, the typical instrument flight in the simulator was to takeoff from Santa Monica Airport and fly to Burbank Airport, where I would execute an ILS approach; then, instead of landing there, I would do a missed approach, and climb up to five thousand five hundred feet. ATC would then give me vectors from Burbank to Van Nuys, where I would execute an NDB approach to a full stop landing. Then, we would re-file an instrument flight from Van Nuys to Santa Monica, where I'd execute a VOR approach to a full stop landing. It sounds like it would only take an hour to do all that, but by the time approach control vectors you all over the sky and you land and taxi back and sit and wait for a tower-to-tower clearance, almost three and a half hours are gone.

•

The first thing Charley discovered about 'trainer type' airplanes, like my Musketeer, was that they're slow. When I flew my airplane, I always used the recommended throttle setting from the Operator's Manual especially for cruise.

The Beechcraft Marketing Department published the performance 'numbers' that seemed to have no correlation with reality. When I flew with Guy and Per, we used the book setting of 2350 rpm for cruise, which was supposed to give a performance figure of one hundred and thirty miles per hour at altitude — but it never did. But since Guy and Per were being paid by the hour, speed didn't seem to matter to them.

But after Charley discovered that the 'book' power setting only gave us a hundred miles per hour cruise speed, he couldn't stand it and told me: 'From now on, the cruise throttle setting is always twenty six hundred rpm. Got it?'

'Got it,' I replied.

'We're in an airplane. And if we don't go as fast as we can, we might as well take the bus!' he said, impatiently. Even wide open, our speed was only around one hundred and fifteen miles per hour at altitude — still shy of the book figures. But any improvement was good that satisfied Charley.

I never questioned Charley about the Musketeer's lack of speed, I figured it was because its fuselage was shaped like a whale, chuncky and wide, with large fixed landing gear hanging down in the breeze on each side and with a large nose gear hanging down in the middle. With that much parasitic drag, it was a wonder the thing could even fly.

But as impatient as Charley was about speed or the lack of it, he was always very patient with me; as patient as any of my former instructors had been, except when I'd continue to make the same mistake more than twice. The German in him would come out, and he'd shout something like, 'if ya touch that setting one more time, I'll cut off both your arms!'

His frustration level at that point was equal only to my own, I kept thinking that I should be doing better than I was. It seemed that the 'ol' plateau' still confronted me. And after a time, even Charley began hinting that I should be flying at least three to four times a week or I wouldn't improve. I finally began understanding that if I didn't fly more often I'd never get any better.

One Sunday Charley made my mind up for me–and it was over. Late in September, I arrived for my usual Sunday morning lesson and a new instructor greeted me at the door. It seemed that Charley's work visa expired after three years, and he had to quit working there. And by the end of the month, he'd have to return to Belgium.

Shock of all shocks! Where had I heard that story before? I felt deserted and forsaken, as if a friend had just slapped me in the face. At least Per had given me a month's notice.

I didn't know what to think. But I knew that my training for the rating was over. My bank book attested to that. There was just enough money left in the account for ten more hours of flying.

By Charley's standards, I reckoned I needed fifteen more hours to 'polish' my skills for the checkride. Not only that, but I would need seven days off in a row to concentrate on the 'polishing.'

Again, it was the same old story of time and money. Fifteen hours and seven days off. Even if I could afford seven days off, there no assurance that it would be sufficient to polish me for the checkride.

There really wasn't any point in continuing to burn what little money remained. I wasn't any better off financially than any other 'working stiff.' I had bills to pay and never enough money to pay them. Whatever money was still in the account would be gone soon enough, once my bill drawer was empty.

•

I was now over five thousand dollars lighter and still no rating. When it was all said and done, I had completed what I started out to do; I now had an instrument scan good enough to save myself in a cloud. But I don't think I needed one hundred and six hours of instrument time just to do that.

This 1969 Piper Cherokee 140 is the same type of airplane the author flew in Chapter 2. Fred Weick, the designer of the Ercoupe, also was the Chief Designer at Piper Aircraft, some years later, and designed the Cherokee Series of aircraft to compete with Cessna's 172 series of aircraft.

The authors oldest daughter, Kristin Pohlmann, is seen posing by a modified 415-G model Ercoupe sliding canopy and 90 horse power engine.

Here is a typical 415-C Ercoupe with its twin tails, 2 degree offset engine and 2-control system that was 30 years before its time. Today's modern high performance single engine airplanes all use inter-connected controls.

The author's beloved 1950 Piper Pacer, that spent the first 30 years of life as a float plane flying the bush out of Seward, Alaska. It was Pohlmann's first airplane ownership, after renting airplanes for over ten years. He owned it for six years and logged over 450 hours of fun travel and adventure with 510 landings without coming close to even one ground-loop. He owes this fact to the great tailwheel training he received from his aviation mentor and long time friend, Bill Hare.

A gathering of 'Short Wing Piper Club' flyers at a flyin at Lake Isabella, California nestled in the Tehachapi Mountains at the four thousand foot elevation. This was in the summer; in the winter the level of the lake makes almost half the length of the runway disappear under ten feet of water. The lake is 26 miles east of Bakersfield, a large cotton producing city in the San Juaquin Valley.

The authors wife Barbara and youngest daughter Lara stand by the Beechcraft Musketeer that Bill and Barbara travelled in during their Mexican excursion of Chapter 7. The plane is parked on the ramp of America's Flyers International Flight School at Santa Monica Airport.

Here the author is shown at the recent Experimental Aircraft Association's Copper State Flyin, at Mesa, Arizona.

The author's favorite cousin, John Pohlmann, poses in front of his beloved Pitts Special as they wait to compete at a Northern California aerobatic contest.

Here Bill Hare, the author's aviation mentor and long time friend, gingerly climbs into the cockpit of the Bowers 1A Flybaby that was built from plans and took two pilots fourteen years to construct.

Bill Hare taking the author's Flybaby out for its first flight. If you look hard, Bill Hare's Red and White Pitts special S-2B can barely be seen in the darkness of the hanger the two Bill's share, at Hawthorne Airport just four miles east of LAX.

An inside view of the Flybaby's spartan cockpit. Note the Pitts Special windshield. Pohlmann states that flying the Flybaby is like riding a motorcycle in the sky, except your teeth don't get so many bugs.

Here the author is shown using the 'Armstrong Method' to prop his custom-built Flybaby, since it has no starter motor.

Here is the author's longtime friend and 'Flying Buddy,' Les Benis (chapter 3 and 4). Les has not only filled a dozen log books with 'powered flight time,' but has filled almost as many for gliders. He has been an International Glider Instructor for most of his adult life. Here he readies his newly acquired sailplane for another adventure in the sky.

Chapter Six
The Thrill of Mountain Flying

Surviving Danger In a Vintage Airplane

There's an old saying in aviation, that 'if you fly long enough, sooner or later you're going to crash an airplane.' Of course, I never really believed that; but at the most unexpected time and place, I suddenly found myself involved in an airplane crash as a passenger in a friend's Piper Pacer. And, of course, the mechanical problems leading up to it didn't manifest themselves until we were airborne and about to fly over the Rocky Mountains. Which validates the corollary to Murphy's Law perfectly. 'When something does go wrong, it will go wrong at the worst possible time and place.'

John Kieth and I became friends after being work associates. Every Monday morning we would sit around the office and do a little 'hangar flying,' as I'd tell him about my weekend flying adventures. John listened with a smile on his face and a far away look in his eyes. On numerous occasions he told me that his dream was to someday become a flyer himself.

I remember one Monday morning in particular, when during our hangar flying, John became adamant about how badly he wanted to fly and own his own airplane. When the tone in his voice turned serious, I finally told him what Les had told me years before, when I dreamed aloud about flying and becoming a pilot.

'John, each year ya get older — so if you're ever goin' to become a pilot before ya die, when're ya goin' to start?. If ya wait too long, life will pass ya by, and ya'll never know what it is like to see what a pilot sees or go where a pilot goes.'

I could see his eyes narrow and his head nodding. To lessen the terseness of the challenge, I also went on to explain how he could actually afford to own a vintage airplane like mine, for the price of a modest used car. And, in addition, he could enjoy the benefit of airplane ownership while taking flying lessons.

Once a person owns his own airplane, the hourly costs for flying lessons are just the cost of fuel for the airplane and the cost of an instructor. Concluding my tutorial, I mentioned that I even knew of an instructor who could take him through to his Private License in a tail dragger — one of the rare CFI's that was 'tail wheel current' in the entire Los Angeles basin.

When I mentioned Bill Hare's name and told John that Bill was the same instructor that checked me out in my Pacer, the tutorial stacked up as all pluses, and no minuses.

At the time, I wasn't sure he'd ever follow through with the plan I discussed, since people usually like to talk about their dreams, but hardly ever follow through with them. But John was different. John was a 'mountain man' from Colorado, and when a mountain man finally gets serious about something–action follows.

I didn't discover that he'd taken me seriously until a few weeks later, when he came to work one morning and calmly announced that he'd just bought an airplane — a Piper Pacer, like mine.

'Well, it's not quite like yours,' he added. It's not a straight Pacer. Originally, it was a TriPacer, modified back to a tail dragger from one with a nose gear.'

Now, to a purist, the only difference between my straight Pacer, model PA-20, and John's airplane, model PA-22/20, was the shape of the side windows of the cabin; the rest of the airframe was identical. On a PA-20, they were oval, while on a TriPacer, PA-22, they were square. So when a Pacer's side windows are square, it means that it began its life as a TriPacer, before being converted to the model of its predecessor, and designated a PA-22/20.

Once I learned about John's airplane, I immediately introduced him to Bill Hare. And by the time they compared their mutually busy schedules, John had acquired a tie-down spot at Hawthorne Airport, four spots down the ramp from mine. And, as things turned out, Bill Hare became John's flight instructor, and six months later John had his Pilot's license.

Once John had his license, we'd bump into each other on the weekends. As one of us taxied in or out of the parking area, we'd pass each other and wave. And at the end of the day, we'd do a little 'hangar flying' on the ramp.

•

John led a really interesting life — as interesting as his appearance. He was five feet ten inches tall, medium build with balding red hair and a full red beard. I think the beard was intended to lend credibility to his image as a scientist. And short of wearing a white laboratory coat, he came pretty close to the Hollywood image, although I don't think he ever spent any time at 'the bench' slaving over a hot Bunsen Burner. The final touch of academia was his black horn rim glasses and white pocket protector that made him look like every man's 'Dr. Science.'

John was a technical consultant, who lived and worked for a company in Colorado Springs, Colorado, but who spent half his time on site working in Los Angeles. I first met him while we worked on a major aerospace program where he was under contract. His schedule was such, that he worked two or three weeks in Los Angeles and then returned to his home company in Colorado for two or three weeks. By the time I'd met him, John had been bouncing back and forth on this crazy schedule for more than five years; and from the looks of things, it would continue for the foreseeable future.

It so happened that John's company paid its senior employees their per diem in one lump sum for the entire year, at the beginning of every January. The single payment covered a year's cost of living at out of state locations where the employees were working. And what John had been doing over the years with his annual per diem check, was just putting it in the bank and drawing against it throughout the year for his Los Angeles expenses. At the time, twelve months of per diem for Los Angeles amounted to fifty thousand dollars.

It was in December of the year when John earned his Pilot's license, that Frieda, John's wife, finally decided that in Los Angeles what they needed to buy was a permanent residence, instead of throwing away the per diem on rent. John's company didn't care how he spent his expense money, so, Frieda thought that they could buy a place in California. That way, after the job was over, they would own a place where their kids could come and vacation, and wouldn't have to spend money on a hotel. And since their kids were mostly grown up by now, that idea made real sense.

So once Frieda had decided, John, in typical 'mountain man' style, just went out one afternoon and bought a place, the way a person goes down to the corner store and buys a quart of milk.

To appreciate the mentality of a mountain man, you really have to know one personally, otherwise, you'd never understand his motivation. Mountain men contain more than their fair share of self esteem, self reliance, tenacity, and the continuing belief that every situation in life is to be taken on, as a personal challenge.

When John said he was shopping around for a place, I figured that the only thing he could buy in the Los Angeles area for fifty thousand was a 'fixer upper' garage. At the very most, I thought he could buy a modest condominium–but even condos cost a lot more than fifty thousand, even in the wilderness fifty miles outside of Los Angeles.

Back at work, our office staff took bets as to what John would end up with, and were totally surprised the very next week when he announced that instead of buying a modest condo or house, he bought a boat.

Disappointed and confused, I replied, 'That's nice John. Did it come with a trailer? Everyone new to California always buys a ski boat with a trailer.'

'Trailer hell,' he laughed. 'It's a big boat! Forty three feet long. I'm going to live on it,' he said, proudly.

'You're joking, right?' I laughed, waiting for him to nod in agreement.

'Hey. I'm not kidding. It's a live-aboard, and it's down at Alamitos Bay, at Marina Pacifica. It's a 1940 Stevens, with a wooden hull. It's a cabin cruiser,' he went on. 'It's even got radar, VHF radio and twin V-8 Chrysler Marine engines.'

'John, that's a Helluva big boat!' I observed. 'No grass to mow, no yard work, That sounds great!'

'That's the whole idea. Frieda and I can live on it when I'm in Los Angeles.'

And that's how John came to live on a boat when he worked in Los Angeles, every other month.

I asked John why he didn't buy a small house or lease an apartment. The mountain man's answer, was — because he'd lived his entire life in Colorado, a state not known for its waterways, and since childhood he'd always dreamed of living on a boat.

Living in Colorado he'd never bothered to learn to swim, especially since there wasn't enough water in the whole state to drown a mouse. With classic logic, he thought that living on the water would be a good challenge for him to take swimming lessons.

To live on the water without being able to swim was something that could appeal only to a mountain man. And from observing John, I've also discovered that if there is more than one way of doing something, an easy way and a hard way, a mountain man will never choose the easy way.

•

Every Spring, John flew his airplane to Colorado from California, where it spent the winter months. And around Halloween, he again would fly it to California before the coming of winter. But, after performing this ritual for several years, he realized that when he was home in Colorado during the summer, he never got to fly very much; he was too busy catching up on chores around his homestead. So he decided once and for all, to move his airplane back to California where he'd keep it for the duration.

Now after listening to John's account of his flights back and forth between Colorado and California, I thought that it would really be fun to ride along with him and cross the Rockies in a small airplane, especially now that this would be his last trip for a while. So after several discussions with him about how badly

I really wanted to go, he gave in, not understanding why I thought it was such a big deal to cross the Rockies. To him the Rockies were just something that had to be crossed anytime you wanted to fly west.

I was really excited about the trip. It would be my only chance to fly the Rockies and see the Grand Canyon from the high plateau country, and I thought it should be safe enough, since John had already flown back and forth a half dozen times. But, I think the reason he decided to let me ride along was that I had more time in a Pacer than he did, by a couple of hundred hours, and offered his trip an extra margin of safety.

The time went by slowly in my impatience for the trip to get underway. The week before the trip, John and Frieda returned to Colorado by commercial air, and the plan was for me to arrive there on Saturday noon by airliner, when John would pick me up and take me to his hangar at Meadowlake Airport, northeast of Colorado Springs.

The plan was for John and I to spend the afternoon getting the Pacer ready for the trip back to California. And since John hadn't flown in a while and wasn't current, he intended to fly around the 'patch' for three touch and goes, to regain his currency While he practiced his landings, I would watch from the restaurant across the runway. This was to happen Saturday afternoon, so that early Sunday morning he and I could just climb in the Pacer and take off.

So Saturday morning found me in the 'smoker' climbing to altitude out of LAX bound for Colorado. And three hours later, I was wandering around the terminal building at the Colorado Springs Municipal Airport, not knowing where to find John. As if on cue, he suddenly rose up in front of me on the escalator from the floor below. Our meeting was so spontaneous and accidental, we both laughed when we saw each other

After a twenty minute drive, we stood in front of John's

hangar staring at his Pacer. We stood by his airplane on a clear and sunny day, with the temperature in the sixties. But, an hour later, as we busied ourselves puttering on his airplane, a fast moving cold front came through and the temperatures dropped into the forties with a sudden cloudburst that lasted a half hour, before changing to hail. Ten minutes later it stopped hailing, the wind came up, blowing the clouds away, and bringing back the sunshine with temperatures back to the high fifties.

I was wearing only a light jacket, and I stood shivering in the cold hanger. I spent the remainder of the afternoon trying to get warm. Here I was again playing the naïve Pilgrim For some reason I always considered myself an experienced traveler, having spent time in the military overseas. But then I recalled the day I spent freezing in Texas, waiting for Les and his Cadet, and now I was spending the day freezing in Falcon, Colorado. Some things never change.

I spent the night at John's house and the next morning, the three of us got up very early. Even by the time we'd dressed and driven halfway to Meadowlake Airport, stopping at an all night restaurant, first light was still an hour away. We each had a hearty breakfast and dawdled with our food, so that dawn would come up before we got to the hangar. Finally, daylight arrived as we drove the last few miles to Meadowlake Airport.

•

John and I had spent last evening planning the flight on his dining room table, so there wasn't much left to do at the hangar, except to pack the airplane and leave.

John figured that during the flight, I would back him up on the navigation, so I brought my lap desk, which was completely outfitted with the latest sectionals as well as a current radio navigation atlas of the entire United States, on which I'd marked our routing with a yellow 'high lighter.' So, as a navigator, I was armed 'for bear.' and ready.

After parking the car, Frieda and I stayed out of John's way, as he opened the hangar, pulled out the Pacer and began loading our gear. Frieda and I vainly tried to warm ourselves from the chilly morning while watching John. No matter what I did, I couldn't get warm, first standing on one foot and then the other. My thin wind breaker was useless against temperatures in the forties, even after the sun's rays finally shown over the eastern horizon.

Finally, John announced that the airplane was loaded, He closed and locked the hangar door. He walked to Frieda and kissed her goodby and told her he'd call her from California. Then he nodded for me to climb into the airplane. It took us a long time to get situated in the Pacer. And after we settled down in the bucket seats, we organized all of our 'flying tools' and got out the various checklists and navigation charts for the trip. We took our time, and as the minutes passed I suddenly realized that Frieda was still patiently standing a few feet away in the cold, stoically waiting for us to start the engine and get on our way. She really was a good sport.

Finally we completed our cockpit organization and John waved a final goodby to Frieda, who returned his wave, and then turned and walked to her car without looking back and immediately drove off.

The Pacer's starter slowly began cranking, and as it moved we could see the propeller turning clockwise through the windshield. It was obvious that the engine was cold from sitting all night, and it wasn't until the ninth blade that the engine reluctantly coughed and came to life.

'Let's wait for the oil temperature to warm up a little; maybe till the needle moves off the peg and into the yellow,'

John said.

I nodded, impressed to think that John would be patient enough to let the oil circulate. Too many pilots neglect to warm up the engine before 'blasting off.' The wear on cold engine parts is tremendous.

Finally, when the needle barely began to move off of its peg, John slowly taxied from the hangar to the taxiway, and then slowly up the taxiway for the full length of the runway, to the engine runup area at the far end of the field.

Sitting in the runup area with the engine idling, John took his time setting up each unit of avionics. During the magneto check of the engine, everything sounded smooth and strong, and by the time John completed the last item on the list, he looked over at me and smiled.

'Ready?'

'Let's go,' I said, giving him a 'thumbs up.'

As a flight crew, we were going to operate using airline procedures; John as Pilot in command would concentrate on flying the airplane, while I would take care of the co-pilot duties, such as operating all of the radios, radio navigation equipment, and the monitoring of the engine instruments. That way, each of us had enough to do during the flight to keep from being bored, but still had enough time to occasionally look out at the scenery.

Actually, we didn't expect much in the way of enroute air traffic, since it was Sunday morning and our routing would take us south from Colorado Springs to a point where we could cross the Rockies at about thirteen thousand five hundred feet. The pass through the mountains was at the end of a one hundred-mile corridor. From there we would pass Alamosa and then on to our first fuel stop at Farmington, New Mexico. I was certain that we'd be hard pressed to see another airplane flying during our trip.

We planned two stops for fuel; one at Farmington and

another at Kingman, Arizona. If head winds slowed us too much, then we'd also stop at either Barstow or Corona just prior to entering the Los Angeles basin.

So, as we pulled onto the runway at Meadowlake, John looked over at me as he pushed the throttle all the way in, grinning, and said, 'Ready or not, we're outta here.'

•

Meadowlake Airport is a modest uncontrolled country airfield, fifteen miles northwest of Colorado Springs, at an elevation of six thousand eight hundred feet. Like all airports throughout Colorado, whether it's summer or winter, they're all considered 'high density altitude' airports. That means, because of the elevation, an airplane taking off from the runway thinks it's already at sixty eight hundred feet in the air. And, at that altitude, performance drops off quickly from what it is at sea level.

And because of its elevation, pilots flying out of Meadowlake must ensure that everything on their airplane is working properly, or takeoffs must be aborted early. At Meadowlake, running off the end of the runway is not something one wants to do. Beyond the end of the runway there is an arroyo two hundred feet wide and fifty feet deep, lying in wait to devour the pilot who misjudges his airplane's performance capability.

The single runway airstrip is forty one hundred and fifty feet long, and thirty feet wide. It's fifty feet lower in elevation at its southwestern end, sloping downhill.

So with high hopes of an enjoyable flight, John shoved the throttle forward and locked it, standing on the brakes until the engine and tachometer wound up to their maximum static values, and then released the brakes. The Pacer lurched forward and picked up speed quickly as it raced down the runway.

At brake release, I punched the 'start' button on the stop watch affixed to the co-pilot's control wheel, using it as one of the indicators of a healthy takeoff: in sixteen seconds at sea level most small airplanes are airborne. So, allowing an additional five seconds for higher altitude takeoffs, I thought that if we weren't airborne by twenty one seconds, we'd better abort and hit the brakes, before running out of airport.

As the airplane began to roll, my eyes were glued to the engine gauges: the tachometer, oil pressure and temperature. In my role as co-pilot my duties on takeoff were to monitor the engine's health, since the takeoff mode is the most critical of all phases of a flight. The engine is hottest, working hardest, with little ram air cooling for its cylinders. If anything mechanical is going to go wrong, it's probably going to happen during takeoff–especially since the airplane is low and slow with little chance to recover if the engine quits.

As we rolled faster and faster, I glanced every few seconds at the digital stop watch in front of me, and then beyond the windshield towards the remaining runway. As the seconds passed 'seventeen, then eighteen,' I watched the runway shorten as we passed the mid-field point.

I looked over at John whose eyes were glued to the airspeed indicator. I glanced over at it and then back to the stopwatch. We still hadn't reached flying speed. The engine was wide open and sounded strong. The tachometer needle was nearing twenty six hundred rpm, as close to maximum power as could be generated before becoming airborne.

At twenty seconds I looked over at John, and he glanced back at me.

'We should be flying by now,' I said.

'I know,' he returned, puzzled. 'It's never taken this long before.'

The airplane continued rolling down the macadam runway with its wheels still solidly on the ground.

'BAM, BAM!'

It was the sound of two quick reports that were heard above the roaring engine; like gun shots or tires blowing out or the twanging of a cable breaking under great tension and hitting the under side of the fuselage. For us to have heard the sound above the engine while wearing our 'hard headsets' meant that whatever the cause, it must be serious.

An instant later, the airplane veered suddenly off the runway at a thirty-degree angle to the right, as it bumped and jostled over the rough infield of the airport. The oblique veering motion shocked both of us, so that neither of us could react.

Immediately, John aborted the takeoff run, pulling the throttle to 'idle' and fighting with the controls to keep the airplane from ground looping. We bumped along over low moguls and knee high weeds of the infield. It wasn't until we were several hundred feet from the runway, that the Pacer finally came to a stop within fifty feet of the perimeter fence.

We just sat there in silence, each wondering exactly what'd just occurred, but thankful that the airplane was still intact. We were three quarters of the way down the runway and never really close to becoming airborne, even in ground effect. For some reason our speed never got above fifty miles per hour. After sitting there a long moment, John finally pulled out the Mixture Control, shutting down the idling engine.

I took a deep breath to relax and pulled the headset off my head. I was confused as to what'd just happened. Slowly, I opened the cabin door, sliding out and into a thick bed of weeds. My body was suddenly impervious to the early morning chill, as I backed slowly away from the Pacer, staring at the bottom mold line of its fuselage, and checking the area around the main landing gear for damage.

'I can't see anything wrong!' I yelled, as John slowly climbed out his side of the cabin. 'The tail wheel looks okay and the rudder cables seem to be intact.'

'Dammit! John exclaimed, almost under his breath. 'What the hell is wrong now?'

'I thought that maybe the rudder cables let go and twanged against the fuselage. It was pretty hard to tell just what kind of noise it was, or where it came from exactly,' I offered. 'What do ya think caused the sideways motion?'

'I think it was just me,' John replied. 'I think I was so surprised by the noise and the fact that the airplane wasn't performing, that I just let my feet relax on the rudder pedals, and the airplane just ran off the runway.'

'Maybe.'

John crawled around underneath the main landing gear, looking at the radius strut that keeps the landing gear from collapsing on itself.

'The rubber donuts look okay to me,' he yelled, from underneath.

The main landing gear were under tension from two donut shaped shock absorber bands that flexed when the main gear would hit a bump, taking up the shock. The radius struts held the landing gear in position and acted as a stop for the shock absorbers. If something happened to them, then the gear would be pulled inward by the tension of five hundred rubber bands that comprised each donut. That was shock absorber technology from the J-3 Cub days of the forties.

'What's wrong?' I demanded.

'Beats the hell outta me!' John answered. 'Everything up front looks normal.'

'Dammit, John! I know I heard something, and it had to be pretty loud to be heard above the engine.'

'I heard it, too,' John agreed, crawling out from under the

fuselage.

I looked at the ground and shook my head. 'Nothing like this ever happened to me in my Pacer.'

John didn't answer, but moved slowly and deliberately around the airframe, closely examining every inch. He spent a lot of time making certain that the aileron cables were still connected. And when he was finished, he stopped near the propeller and stared down at the ground, jamming his hands into the pocket of his flight jacket.

'I really have no idea of what made the noise. I can't even figure out why the airplane wasn't performing. It's not even grossed out! We're a hundred and fifty pounds under max gross.'

And then, after a long moment, he looked over at me and asked what I thought we should do. Personally, I was ready to throw in the towel and thought we should taxi his airplane back to the hangar and let his mechanic go over it in detail. But, John was a 'mountain man' and macho peer pressure wouldn't allow him to 'whimp out' of a trip which both of us had invested so much time and money. The one way airline fare to the 'Springs had cost me over two hundred dollars, and we both knew that John's work schedule wouldn't allow another window of opportunity like this, for almost another half year. So it was really now or never.

I knew exactly what he was going to say, so I just silently shrugged and waited for him to say it.

'Hell, I say we taxi her back and try it again!' he laughed. 'Ya know, sometimes these things never do get explained.'

I nodded, knowing full well that something was terribly wrong. I just couldn't find any evidence of what that 'something' was, and the thought that we couldn't pin point the problem made me real uncomfortable. I had the feeling that John was real uncomfortable too, he just wasn't talking about it.

With John on one side and me on the other, we took hold of the forward wing strut and began pushing the airplane back towards the runway. Laboriously the Pacer moved through the thick weeds, and after several minutes was again sitting on the runway.

'We'd better get her off the active before somebody flies in and lands on top of us,' I said, jokingly.

'Come on,' John yelled, as he climbed into the cabin from his side. And before I could get back into the airplane and buckle my seat belt, John had the engine started and after a moment the Pacer was rolling slowly along the asphalt, down the remaining runway to the turnoff for the parallel taxiway.

Once on the taxiway, it took us a long time, at two miles per hour, to taxi the full length of the runway back to the engine runup area.

As we made the turnoff for the parallel taxiway, I couldn't help glancing over at the arroyo and hoping that this time, if we did take off, we'd at least get airborne with enough altitude to clear the crevasse.

When we got to the runup area, John shut down the engine got out, opened the engine compartment, and for what seemed a very long time, busied himself checking every conceivable component that his hands could touch, forward of the fire wall. I guess he was making certain that there wasn't anything loose, wet, or dripping.

But, as far as I was concerned, I already knew that he wouldn't find anything out of the ordinary. I could tell by the sound of the engine that it was smooth and strong and putting out every ounce of horsepower it was capable of. The engine had nothing at all to do with the noise we heard, but I didn't say anything to John, and just sat there in the cabin staring down at the radio navigation chart in my lap.

John continued checking for a few more minutes. It was

obvious to me that if we hadn't found anything by now, then it was something that wasn't visible to the eye.

While John satisfied himself about the Pacer's engine, the realization came to me that our situation was a classic case of two flyers not talking to each other and sharing their true feelings. Flying magazines wrote articles about this very thing entitled: 'Never Again,' or 'I Learned From That.' And, upon reflection, what we should've done is cut our loses and go back to California without the Pacer. It was evident the airplane had a problem that wasn't going to be easy to fix.

Peer pressure was surely motivating us to press on at any cost, a strategy that was not only stupid, but dangerous. There was no getting around it, the flight to California was probably doomed, and should be scrubbed. That would have been the smart thing to do, and then we could 'call it a day,' and it wouldn't be anybody's fault and nothing serious would happen to us. The signs were all there; all we had to do was heed them.

But being educated and sophisticated, not to mention two men trying to show each other how unafraid they were of fate, we pressed on, all the time, wondering if we were really going to 'meet our destiny.'

When a flyer refuses to fly because of a safety issue, that's okay — that's something you can point to and say, "I'm not going to jeopardize my life." Other pilots will still respect you. But, if you refuse to fly because of an undefined noise, that's another story. No one could really trust you as a flyer after that.

John was a scientist, and as a class, scientist don't believe in magic, so if there's no physical evidence of an airplane's inability to fly, then there's nothing wrong, and it should be airworthy.

As I thought about the dynamics of the situation, I began to feel a very small ache in the pit of my stomach. I'd felt it before. It was a warning sign that things weren't right; and that we should beware. It wasn't as if I was nervous or afraid, not yet.

But the ache persisted, and from years of flying into situations over my head, I know what that ache meant. It was the beginning of fear. If it persisted, it would grew into a stabbing pain. But for now, it was just a small ache.

Although he wouldn't say so, I think John was worried about all of the money I'd spent just getting to this point of the trip, and I think that he was afraid that I would be disappointed if the flight was aborted. And, maybe if we'd aborted the flight right then I might've been disappointed; but, so what! I've been disappointed before, and sometime in the future, I'll be disappointed again. The important point here is that John and I should have talked to each other about what we really felt, but we didn't.

I just let John finish checking the engine. And finding everything in order, he would naturally continue the trip. After all, he was a scientist, and scientists react to situations predictably. If there's no evidence of a problem, then there's no problem.'

Problem or not, the airplane still didn't get off the ground with its engine running at full speed. Secretly, my hunch was that the trouble had something to do with the rubber donuts; after all, the temperature last night had been below freezing and rubber has a way of getting brittle when it gets too cold. And if the rubber bands got brittle, it wouldn't take much for them to snap in half. The more I thought about it, the more it seemed plausible; the radius strut would continue to keep the gear erect, even if one of the main gear shock absorbers had broken.

The fact that I'd mentally worked out the only scenario accounting for the lack of visual evidence was of no consequence, and I never verbalized it. No scientist could be bothered with an unsubstantiated hypothesis.

John proclaimed the engine 'operational' as he climbed back into the cabin, and shrugged as he glanced at me fastening his seat belt. I returned his shrug with a nod, while he again took out his checklist and offered it to me.

Dutifully, like co-pilots everywhere, I read each item out loud over the intercom without looking over at John, and he responded over the intercom with a terse 'check,' or 'normal,' as if this litany would change the final result of our next attempt at flight.

After the engine started we waited for several minutes while the oil again could warm up, moving the needle into the middle of the 'green.' This time when John did the magneto check, he ran the engine up to its full static power which the needle of the tachometer verified.

After the runup was finished and the engine was again idling, I looked over to John and said, 'It's a good thing Frieda didn't stick around to watch us abort that takeoff, or she'd probably be worried.'

'Ya,' John returned. 'I thought about that. I'm glad she's not here, cause I know she'd be worried.'

•

So once again we taxied onto the runway, but this time there was no hesitation, and John immediately pushed the throttle forward as I began another blind radio broadcast of our intentions to whatever traffic was in the area. This time as we rolled faster and faster I could see John watching the oil pressure gauge and tachometer very carefully for any fluctuations.

The actual takeoff began as the Pacer was halfway down the runway. As if in slow motion, the airplane broke ground and continued its climb out. It was dreamlike as the airplane lifted off but John's gaze was focused straight ahead at the runway that was slipping below our wheels and out of sight.

I wasn't sure that we were ever going to leave the ground, as the engine droned on at full power. I didn't bother timing the takeoff, as the outcome would be evident soon enough. John horsed the Pacer off the runway at the last second and the airplane staggered into the air below flying speed. We were up, but we weren't flying. The wheels were two feet above the runway and the airplane was flying in 'ground effect,' a cushion of air that would soon disappear, once we ran out of runway and airport. We had to increase our airspeed so that we could sustain lift over the wings without a buoying effect from below.

As the last of the runway moved under our wheels, I looked up ahead at the growing view of the arroyo.

'Drop the nose. We've gotta pickup flying speed!' I yelled over the intercom.

John nodded, staring straight ahead with his hands in a 'death grip' on the control wheel. Very gently he lowered the nose just five degrees towards the horizon. The movement of the control wheel was almost imperceptible; — a movement akin to those used in 'slow flight,' less the airplane would stall.

I looked over at the airspeed. Was it enough of an attitude change to pick up the speed we needed to sustain flight on our own?

The airspeed needle pointed to '85.' That was our best rate of climb speed. And although we were now flying on our own, we didn't seem to be able to climb.

'John, we've gotta climb. Make it climb!'

Desperate, John 'horsed' the control wheel back towards his lap, forcing the Pacer's nose another ten degrees above the horizon, his arms struggling against great control pressures. But, after a few seconds at the new attitude, the vertical speed needle suddenly came to life, climbing off the number zero. First, twenty feet per minute, then thirty, and finally forty. It seemed to be stuck on the number forty. But, forty feet per minute was better than nothing.

Normally at altitude, one hundred feet per minute or less is considered an airplane's service ceiling, where it's about as high as it can fly. The problem was, we weren't anywhere near the Pacer's service ceiling, but something was causing it to act as if it was.

As the needle reached the number 'forty,' the arroyo flashed by below us, as the airplane continued a labored climb. It wasn't on the verge of a stall, but it seemed to be shaking a little, as if we were asking too much of it.

Outside, the temperature was still cool, and even at our present altitude we should've been able to climb at least at a rate four times what we were, Even with full fuel, at our current weight the airplane should be flying much faster. The same 'problem' that caused us not to take off on the first attempt was still with us. It was holding us back, keeping the airplane from reaching full speed.

After several minutes the climb rate was one hundred and ten. The wings were acting like they were flying at fifteen thousand, instead of seven.

I glanced at John's hands. He continued to hold the control wheel as if squeezing it would make the airplane climb faster. After several seconds, when it appeared that the Pacer would stay in the sky, I glanced at John and said, 'Ya better trim off the pressure before your arms get rigor mortis.'

'You're right,' he smiled, turning the trim crank a half of turn.

I looked at my watch. We'd been airborne for more than four minutes and yet, we weren't even five hundred feet above the ground. I looked at the chart in my lap. We were quickly approaching the Airport Traffic Area of Colorado Springs Municipal; an airspace we couldn't penetrate without prior permission.

'Ya know, John,' I said clinically. 'I don't think at our present climb rate we're going to clear the ATA.'

John glanced over at me and shrugged.

'What's wrong? Can't we climb any faster?'

'I don't know what's wrong,' he returned, moving the trim wheel again. 'It's taking a hell of a lot of pressure just to keep the nose up!'

Ten minutes later and ten miles closer to the Municipal Airport, our altitude was still only two thousand feet above ground level.

'I've gotta call the Tower and get permission to transit their airspace before we bust it,' I said, wondering to myself why we weren't thinking about aborting the flight altogether.

'Okay,' he agreed. 'I don't know why we can't climb any better. There's a hellava lot of parasitic drag–I don't know what's causing it.'

First, I called the Tower at Colorado Springs and got permission to transit their airspace; then I called Flight Service at Denver and reported our plight and canceled our flight plan

The engine droned on at full power for another ten minutes. We were now thirty miles south of the 'Springs, where we were to turn westward and enter the hundred mile corridor that led to the pass that would take us through the Rocky Mountains.

This was the part of the trip that I'd looked forward to, one of the main reasons that I'd tagged along. But now, with this drag problem holding us back, I wasn't so sure that we should even be attempting to fly through the pass, let alone continuing the trip. We were barely maintaining flight just twenty miles per hour above a stall.

The pass was at the end of a hundred mile long valley, of which ninety miles had to be traveled before one could see the pass itself and determine whether or not it was open 'weatherwise.' What would make the pass impassable was thunderstorms or snow storms.

If the pass wasn't open, then the navigation problem became fuel management. If we couldn't continue through the mountains towards Farmington, we'd have to backtrack the entire hundred miles to the point where we first entered the mountains, burning up two hours of fuel for nothing. Then we'd have to travel to the southern tip of the Rockies, where they petered out near Las Vegas, New Mexico, and then head for Albuquerque.

There was no way we could make New Mexico at the rate we were consuming fuel.

We were now at thirteen thousand five hundred feet, and unable to climb any higher. Five minutes went by, then ten. I looked at my watch; we'd been airborne for an hour, and the fuel gauges told me what I feared would happen; that the left tank, the one in use, was getting pretty low–it probably didn't have more than twenty more minutes of fuel.

'John,' I said over the intercom. 'Look at your left fuel gauge. We've gotta switch tanks before it runs dry and kills the engine.'

Surprised he glanced over to me. 'I don't understand? That tank should have a lot more fuel in it.'

'I know, but think about it,' I observed. 'First, we've been running the engine at full power ever since we took off, an hour ago. And, second, leaning the engine hadn't saved much fuel, since the engine was pulling us through the air despite our drag problem. Plus, our airspeed has been only eighty five miles per hour. We've been using fuel at an alarming rate.'

'Jesus!' he muttered. 'Ya know, usually on this trip I get almost three hours out of each tank, and now it's almost empty.'

As he switched tanks I wrote down the time.

'If the pass is open, we'll have enough fuel to make Farmington, all right,' he said, glancing in my direction.

'And if it's closed?'

'Then we'll backtrack and find someplace to refuel.'

'Ya, but at the rate we've been using fuel, we don't have much more than an hour from now,' I said, flatly. 'The drag is killing our range, especially since our ground speed is only about eighty miles per hour!'

He nodded, and sat up in his seat stretching his neck to see something in the distance ahead.

'Oh shit!' came booming over the intercom.

'What's the matter?'

'Look ahead,' he said, motioning with his head, towards our eleven o'clock position. 'The goddam pass is closed. It's snowing like hell there. It must be a blizzard. There's no way in hell we can get through.'

His voice was still ringing in my ears and the ache in my stomach was slowly growing into a pain, as I strained my eyes to see the horizon. Peering in the direction of the pass, all I could see was a fogginess, which must be what a snow storm looks like from miles away.

John moved the controls very gingerly, since the airplane was barely flying above stall speed with very limited control.

John kept the bank angle at five degrees, and after what seemed an eternity, we were finally headed back towards the entrance of the corridor, some ninety miles behind us. It was a strange feeling, being in slow flight. The engine was racing full speed with the engine leaned to the point of running on fumes, and yet the airplane moved as if it we were suspended in space and only turning on a stationary point. But even so, the fuel was being consumed at the rate of fifteen gallons per hour and the tank only held eighteen gallons.

Now that we were headed back, we had nothing else to do for the next hour but crank out the miles. And to pass the time, John and I began what could only be called a totally thorough recheck of every part of the airplane that was accessible to our view, especially control surfaces, as well as anything else we could think of. Somehow we had to determine what was wrong.

The list was short, as we'd checked everything before. There wasn't much that we could see from the cockpit that we hadn't stared at during the last hour. I didn't think we would discover whatever was wrong while flying, but being proactive was better than just sitting there like passengers, hoping that things would get better.

We flew onward in silence. As the miles passed, I mentally reviewed the various scenarios that could account for our plight. But it was a short list, and I kept coming back to the shock absorber donuts. There just didn't seem to be any other answer.

Our airspeed at this altitude should have been about one hundred and thirty miles per hour, like the performance charts promised; but here we were staggering through the sky at eighty five. I finally gave up worrying about it, deciding that a mechanic could discover the problem once we were on the ground at our fuel stop.

●

It took us an hour and twelve minutes to get back to Victor 19, out of the corridor and east of the Rockies; back where we started from, some two hours before.

I looked at the fuel gauge again. 'We don't have enough fuel to make Las Vegas, let alone Albuquerque from here,' I announced over the intercom, staring at the numbers in my flight computer. 'We only have about forty more minutes before Isaac Newton will get us down 'the hard way.'

'I agree,' John acknowledged, looking at his watch after a glance at the fuel gauge. 'Find someplace in northern New

Mexico that has fuel. We're now about forty or fifty miles south of Pueblo on Victor 19.'

I looked down at the flight atlas in my lap and traced our route with my finger from the entrance of the pass to Victor 19, searching a fifty-mile radius for airports with fuel. The problem was that the atlas only showed three airports within that range, but there was no symbol for facilities or fuel.

'John, I can't waste anymore time, I'm calling Center. This damn atlas doesn't show whether an airport has fuel or not.'

I found the radio frequency on the margin of the atlas page, and reset the radio. After waiting several seconds to see if the frequency was clear, and hearing no conversations, I pushed the switch and tried to sound calm.

'Albuquerque Center, Pacer 8752 Yankee, request.'

There was no reply. I waited twenty more seconds, and still nothing. It was Sunday morning and probably a lull period for the airliners. I double checked the frequency setting with the atlas and was about to call again when a very deep and relaxed voice responded over the headset.

'Pacer calling Albuquerque Center, go ahead with your request.'

I was hoping to sound as unemotional and calm as the pilot's in the movies, as they transmit their plight to the tower, with their airplane on fire. But, try as I might, the tenor of my voice exposed my fear and frustration. My throat was squeezed shut, like an invisible hand was choking me. I took a deep breath and keyed the mike.

'This is 52 Yankee, and we're a PA-22/20, at thirteen point five, forty five south of Pueblo, on Victor 19. We've got some kind of aerodynamic drag problem. Our airspeed is seventy five knots at full power. We've got about forty minutes of fuel remaining. We're requesting vectors to the nearest airport with fuel. Over!'

'Understand, 52 Yankee, squawk 0286 and Ident.'

I set the new code into the radar transponder in front of me and pushed the 'Ident' button. A few seconds went by and then the voice returned.

'52 Yankee, radar contact. We're showing you about fifty miles south of Pueblo. The nearest airport with fuel from your present position is Raton, New Mexico.'

'Roger, Albuquerque,' I replied, looking down at the atlas. I found the Raton Municipal airport about thirty miles over the border of New Mexico. 'Could ya give us a vector to Raton?'

'Sure thing,' he answered. 'Turn left to 110 and good luck.'

As soon as he said the heading John began banking the airplane. I glanced at the fuel gauge and decided that we had enough fuel to make it and began to relax a little. And, as we cranked off the miles to Raton, my thoughts wondered where our trip would end and if we'd ever discover what our drag problem was.

•

Raton, New Mexico was an old coal mining town that had been paid out long ago. It sat on a high plateau and had two airports; a small field in the center of town called La Mesa Park, and a very large one ten miles south of town, called Municipal. We decided to land at the larger of the two, hoping that there was a mechanic on duty who could fix whatever was troubling our airplane.

Even at this point of our trip, we still believed that we were going to get to California. Since it was John's airplane, I wasn't going to be the one to throw in the towel. After all, John was a 'mountain man' and would never understand or forgive a quitter.

It was eleven thirty in the morning when we arrived over the town. The streets were deserted. Everyone was either in Church or at home.

We began descending to eight thousand five hundred as soon as we got the vector from Center. The elevation of Raton's airport was six thousand three hundred and forty nine, and as we flew over the field we were just two thousand feet above ground level.

The field was uncontrolled and, as it appeared in front of us, John gave them a call for an advisory on the Unicom frequency. Within seconds a man's voice announced the field's altimeter setting and wind direction and favored runway. And, as an afterthought, asked if we needed fuel. Since John was busy, already going through his pre-landing checklist I responded with a quick 'affirmative.'

The airport had three runways arranged as a triangle. The wind was favoring the longest of the three and that's the one John was paralleling on a downwind leg. From seven thousand feet the runways looked very large, long and wide, the white gleaming cement in stark contrast to the faded brown colors of the high plateau country.

As we turned base leg for Runway 20, I took all of the charts in my lap and stashed them on the rear seat, thinking that John had done a magnificent job of pattern work and had set us up for a text book landing. We crossed the numbers just a few feet above the runway and the main gear touched down a few seconds later. It was a 'greaser' of a landing, the kind that all pilot's wished every landing was like.

What occurred next takes longer to describe than it did to happen. Memory always replays scenes in slow motion, but during the event, it actually seemed to occur that way.

With full flaps deployed, the tires touched down at about sixty miles per hour — stall speed. The rollout began normally until the speed bled off and the weight of the airplane was transferred from the wings to the main landing gear–that's when it happened. That's when the right main landing gear collapsed.

We had slowed to perhaps fifty miles per hour when the right gear folded up under the fuselage, causing the airplane to swerve to the right. John immediately tried to counter the swerve with full opposite brake. But, as soon as he did, the left brake locked up and caused the Pacer to ground loop to the left. Then, in less than half a second, the propeller struck the runway a dozen times, stopping the engine dead.

As the right gear collapsed, the right wing tip dug into the runway and performed two complete cartwheels before skidding backwards for a hundred feet, coming to rest against the edge of the runway between two runway lights, missing the nearest one by a foot.

As soon as the airplane came to rest, we both sat unmoving for a full second, not believing what had happened. Then I quickly unbuckled my seat belt and opened the cabin door, jumping clear of the fuselage. Getting out of the cabin took less than two or three seconds, and, as I stood on the runway staring back at the airplane, John sat frozen at the controls, shock and disbelief paralyzing his body.

'Get out, John. The flight's over.'

Slowly, John began to opened the cabin door and crawl out. He had trouble at first opening the door; the entire cabin's framework of steel tubing had been bent by the stresses of the ground loop. But after a moment of pushing, he was able to horse the door open through the bent door frame.

Because it was a high winger, and the Pacer's wing tanks were up out of reach during the ground loop, fire or explosion was never a threat. It might've been a different story had it been a low winger. Real life airplane crashes aren't quite as dangerous as the airplane crashes in the movies, where if an airplane lands 'wheels up' they always catch fire or explode for no reason.

John emerged slowly from the airplane, like a grizzly bear coming out of a cave after a winter's hibernation. At the time I was unaware of the beating that our spines and hips took in the 'whip lash' of the cartwheeling airplane; but, it would only take another day before I felt like I'd been in a gang fight. Both of our bodies had been thrown violently against the restraints of the seat belts and the sides of the cabin; first one way and then the other, as the airplane cartwheeled and skidded sideways and then backwards, across the tarmac.

Once John was out of the Pacer, we stood together in silence for what seemed like a very long time, as we stared at the wreckage of what was left of John's airplane. One minute it was flying, although just barely, and then the next it was a pile of parts.

As if John and I were High Priests performing an ancient mystical rite, we would stand for several minutes at one spot and quietly survey the damage, and then move several feet further along the wreckage scene, where we would again stop and stare before moving on once more. In that deliberate fashion, we examined every inch of the airframe, as we moved around the remains of the Pacer.

When we got to the propeller, we discovered what damage a cement runway could do in reshaping an aluminum blade spinning at the rate of about twenty two revolutions per second. The force of the propeller on the cement stopped the engine in a half second, and had bent back each propeller tip in a large curl that was about six inches in length. I mentioned to John that a propeller shop should look at it before he arbitrarily junked it. Anyway it was worth a try, as used propellers cost at least a thousand dollars.

The engine would also have to be torn down and the crankshaft examined for damage from the propeller strike. If the engine had to be rebuilt or exchanged for a rebuilt one, the cost could vary between eight and ten thousand dollars; new engines from the factory could run as high as twenty three thousand.

The remainder of the airframe damage was really not obvious — it would take a much closer inspection by someone who knew what they were looking at. If possible, a mechanic could tear down the entire airframe and realigned and reweld it, and then re-skin it with fabric.

One thing was for sure; this was where our trek would end. With the Pacer totally destroyed and lying at our feet, our flight was now over.

During our inspection of the wreckage we discovered that the ground loop had damaged both tips of the horizontal stabilizer, as well as distorting the tubular steel longerons of the fuselage. The damage was complete. We were still looking at the wreckage when a white airport sedan drove up on the taxiway and screeched to a stop.

•

It was the airport manager, and as he told us later, it was he who answered our radio call. He got out of his car and stood next to us staring down at the wreckage and shaking his head. He said that he was waiting for us at the fuel pump after watching us set up for final approach. After fifteen minutes and not seeing us takeoff again, he realized that something serious must have happened, and jumped in his car to investigate.

He understood immediately what happened to the Pacer after seeing the right main landing gear tucked up underneath the fuselage. To help put our minds at ease, he told us just to get our gear out of the airplane and relax while he took care of everything. And what happened next was something that was amazing indeed, and could never have happened in Los Angeles or any other metropolitan airport.

While John and I stood around looking helpless, the following things happened within the next half-hour.

First, the manager reached into his airport car that was parked fifty feet away and pulled out a cellular phone which he used to call the Hertz car rental agency in town. Even though the agent was in church at the time, he was located and delivered a rental car right to us on the runway. When the car pulled up and the agent got out, John signed the rental papers right there on the hood of the rental car, a hundred feet from the Pacer's remains.

Next, since there was a National Guard Armory based within fifty yards of the airport, the manager then phoned the Officer of the Day from his car. Within ten minutes, a large five ton truck arrived complete with a rear mounted derrick, as well as twelve National Guardsmen who happened to be on duty that weekend.

And within a matter of a few minutes, the entire crew had the Pacer suspended several feet above the ground, after attaching several cables around the airframe. The truck slowly drove the wreckage to a tie-down spot near the airport office, where it was deposited until such time that John and his sons could come and rescue it.

Next, the manager had John and I drive the rental car over to the airport office, where we could settle up for all of the help. John paid for a tie down spot, and used the FBO's phone to call the airline ticket counter at Colorado Springs Municipal Airport for flight reservations back to California, later in the day.

While Raton is just over the border from Colorado, it was still two hundred and fifty miles south of the 'Springs and, even driving very fast in the rental car, the distance would take us a good five hours.

When John finally settled up with the Airport Manger, the

total expense incurred was only a one-month's tie down fee. The National Guardsmen wouldn't take any money for their help or the use of their equipment. It was my impression that they weren't about to profit by someone else's misfortune. Of course, if this had happened in Los Angeles, the car and the tow truck and the tie down would have amounted to about four to five hundred dollars, with everyone involved holding out a hand. There's a lot to be said for small town people and their values.

•

The drive back to the 'Springs took us four and a half hours, and by 8 p.m, Los Angeles time, our adventure was over and we were back in California.

The adventure was over and it cost me five hundred dollars in airline fares. All I had to show for my flight over the Rockies was a whiplash of the spine and pelvis that cost me another two hundred dollars in Chiropractic visits.

A month later, John sent his sons down to Raton with a truck and flatbed trailer to pick up the Pacer and to haul it back to Meadowlake Airport where he had it rebuilt over the next two years. John didn't know it at the time, but for the cost of an entire rebuild of the Pacer, he could have bought a really nice vintage Cessna 182.

Half way through the rebuild, John's mechanic announced that the real cause of the accident had been a bad weld on the right main landing gear strut. Apparently the strut had been removed and rewelded four inches further forward, during the modification from a TriPacer back to a Pacer; it was a 'cold weld' that we heard snapping during our first takeoff attempt. It gave way, but didn't break off from the fuselage entirely until we were airborne after the second takeoff. The inflight drag was caused by the dangling gear.

Sometimes unforeseen things happen to the safest of flyers. All a pilot can do is to act in as safe a manner as possible. Attention to detail regarding all phases of flying usually gets the airplane home safely, and if the airplane arrives safely, so does the pilot.

Chapter Seven
Have Wife, Will Travel

Is The Airplane Really a Recreational Vehicle?

Ben Short was aptly named. He stood five feet five-inches tall and weighed in at about one hundred and sixty pounds. In the military he would have been referred to as a 'feather merchant.' He was good looking with thick dark hair and a quick and infectious smile. Add all that to an outgoing personality and you have the image of a Navy jet fighter pilot — something he actually was, for most of his adult life. For once the image and the reality were the same. What you see is not only what you get, it was the real Mc Coy.

Of all his physical attributes, I think his prize-winning smile is what attracted people to him like a magnet. After Ben laid a smile on you, you couldn't help but like him.

Ben spent twenty-five years in the Navy, rising to the rank of Captain, and he flew the Navy inventory of fighters from the mid-fifties through the late seventies.

I first met Ben while we were both working in the Aerospace Industry during the mid-eighties. His demeanor immediately impressed me, but I never got to really know him until several years later, when we both got transferred into the same program. It was then that I learned that he'd been a military flyer, but hadn't flown since his retirement from the service.

Over the months at work, when we'd bump into each other in the hallways or parking lots, we'd do a little hangar flying, — he discovered that I was a private pilot who owned my own airplane. And when a couple of pilots get together, no matter how the conversation starts, it always turns into some hangar flying.

One day, Ben popped into my office and asked how he could get back into the air as a private pilot. It had been twelve years since he'd even seen an airplane, but suddenly he was very anxious to be flying again. Since he had no idea of how to go about getting qualified as a civilian pilot, he asked my advice. He was concerned that the task might be too daunting. So I sat him down and we had a long talk about what he had to do and how much it would probably cost. And since he was anxious to get started, I took him over to Security Aviation at Hawthorne Airport the very next day and introduced him to Bill Hare.

Bill Hare was very impressed with Ben's flight experience and assured Ben that although there were hurdles to be jumped, it would be more of a 'walk in the park.' As Ben and Bill stood there on the flight ramp talking, I realized that I was watching two highly experienced aviators acknowledging the other's flying skills. It only took a couple of minutes for them to size each other up.

The FAA could have easily fixed Ben's problem if he had separated from the active military service within the last ninety days. The FAA would had just converted his military pilot's license to a civilian one with an instrument rating. But, it had been twelve years since Ben had even kicked the tires on an airplane. To get qualified and current today would be almost like a civilian starting from scratch.

Bill explained it to Ben that if he wanted a commercial license, he would have to take the Instrument Rating written examination, and then get checked out in a high performance airplane, like the Cessna Skylane retractable that the FBO had on its rental line. Aviation problems for Bill always had an easy answer, no matter how much work they turned into. Bill told Ben to buy the 'red book' of questions and answers for the Instrument Rating written and just read it through twice and then take the test.

For most people, just to read through the test book twice, — containing one thousand questions with multiple choice answers — would be difficult; and then after reading it to take the four hour exam would not only be difficult, but psychologically traumatic. But Ben was highly motivated and very smart. Besides, how tough could it be — it wasn't like flying in Nam where people were trying to kill you. To Ben, meeting this exam challenge only meant that he was finally back on the road to flying again.

How tough could it be? He had all of the aeronautical knowledge in his head, and all he had to do was to relate it to Civilian flying. After all, the FAA regulations applied to all military flyers when they were flying within CONUS, and he'd been flying in and out of the states for more than twenty years.

For the next several weeks every time I'd go into Ben's office, he was busy studying the red book. And because of his motivation and dedication and focus, it was just a few weeks later that he told me that he'd taken and passed the written and was now beginning the airwork part of the curriculum.

As usual, it turned out that Bill Hare was too busy to take on Ben as an instrument student, but he made certain that the CFII assigned to train Ben was the next best on staff. The next best CFII turned out to be a woman. Bill recommended her as being top notch and very demanding.

When Ben was taking his flying lessons, the FAA, to cover themselves, decided that just because he'd already passed the combination Instrument and Commercial Pilot written test, they'd better require him to pass the Private Pilot written, as well. After all, it wouldn't be logical to issue a Commercial License to someone who'd never gotten a Private License first. There had to be a logical progression and an accountable paper trail that the bureaucrats could defend.

Undaunted, Ben went out and got the red book for the Private Pilot's exam, read through it once, and then took the written test and passed.

And, by the time he'd completed the written exam and received his Private Pilot's License, he was ready to take the checkride for his Instrument Rating. He wasn't sure just how easy that would be. Getting his 'scan' back after twelve years wasn't a trivial matter, although it didn't seem to take as long as he thought.

I stopped by his office during the period of his instrument flying refresher lessons, and listened to him relate how good or bad the lessons were going at that moment.

When it didn't go well, it was usually a matter of his 'getting behind the airplane,' or to put it another way, of letting the airplane 'get ahead of the pilot.' In either case, to a flyer, that problem is always bad news. Once you get behind during an instrument flight, it would be easier to be a 'one-armed paper hanger,' than to play catch up with an airplane, especially when you can't see anything but the instruments in front of you.

But several weeks later, when I talked to him again, he'd just taken the checkride and passed. As he related the flight to me, it sounded like the Examiner put him through the mill. The flight itself took over two and a half-hours. By the time it was over Ben was totally exhausted. He said that it was harder than any Instrument Competency Check he'd ever taken in the Navy.

Although Pat, his instructor, had been very demanding as she readied him for the checkride she picked an FAA Designee Examiner from the Long Beach Airport, a man who she knew was tough but fair. She just didn't realize that since the Examiner had never been a military flyer, Ben would have to prove himself against the most difficult challenges the Examiner could think of.

It turned out that Ben got put through the wringer. After the Examiner realized that Ben could handle the normal instrument workload, the remainder of the checkride was performed with partial panel, including, all of the approaches. In partial panel instrument flying, the more important 'direct reading' flight instruments are covered, forcing the pilot to use secondary instruments that only indirectly indicate the airplane's attitude. It would take a virtuoso to pass such a test. Ben had only flown fifteen hours of instrument refresher flights in preparation for the checkride, and the Examiner couldn't understand how a pilot, after a layoff of twelve years, could pass a checkride with so little preparation. Actually, the Examiner underestimated not only the military flight school's teaching methods, but also Ben's skills.

The flight test consisted of the usual airwork as well as unusual attitude recovery. Then the three approaches were executed, each using a different airport, all performed using partial panel. It was one airport after another with one type of approach after another. And by the time it was over Ben was sweating and tired. It took total concentration on his part to keep up with such a hectic workload. Something that would never be asked of him flying instruments in real life.

The Examiner didn't want to believe that Ben zoomed through the various challenges with better than average performance. He really bounced back after such a long period of inactivity; but then, Ben was highly motivated, and it was obvious that the military flight schools are very effective in their teaching methods. The Navy teaches a man in a way that he never forgets. And I personally suspect that Navy aviation training is superior to that of the Air Force or Army.

The other thing that catapulted Ben back into the sky was his focus and dedication to the task. Ben's hunger to fly consumed him. It was clear to anyone who knew Ben that he was destined to fly again; nothing else mattered to him.

I suppose if the Examiner had been a former military jet jock like Ben, the checkride would have concluded in forty minutes. But a civilian can't appreciate military training methods and techniques. The fact that Ben zoomed through civilian flight like a meteor didn't surprise me at all, after Ben's scan and the rhythm returned. After all, to a high time pilot who'd spent twenty five years flying fighters while people were shooting at him, the instrument competency checkride with no incoming SAMs or triple A or bobbing carrier to land on, was just a whole lot easier.

So after twelve years of being grounded, Ben was back in the air again. And within a month of getting his Commercial license, he bought an airplane: a Piper Arrow. It was after he'd told me that he was looking for his own airplane that I loaned him all of the best consumer publications on used airplanes that I had in my library. The books explained both the good and bad points of each model on the used airplane market from airplanes of the 1940's to the present.

He told me later that he studied the books for several weeks before deciding on the Piper Arrow. And after looking at several Arrows, with various sized engines and avionics suites, he bought the one that had the largest engine: 200 horse power, and had the latest IFR avionics. The thing that clinched it for him was the fact that the airplane was in perfect condition, needing only a new paint job.

Suddenly, Ben was a happy man. He was flying again, and he was a proud owner of his first airplane. The only thing that bothered him about his airplane was its color; it was a faded kelly green. The paint was old and peeling. I told him not to worry about the paint, since airplanes can fly just fine for years before a new paint job would be mandatory, especially in California, where a new paint job would cost five times what it should, because of pollution laws. In California, a fifteen hundred-dollar paint job was now six thousand dollars.

But Ben couldn't stand having an airplane that looked new on the inside and grungy on the outside. I guess he didn't appreciate the protection of 'Urban Camouflage.' He brooded about it for weeks. The longer he thought about it, the more it 'bugged' him. And after another couple of weeks he decided to get it painted.

There were only three available paint shops in the southern California area, but after considerable research, he found some bargain prices at a shop located in a terribly remote location ten miles west of Apple Valley in the Mojave Desert, on a private ranch. The rancher gave up ranching to paint classic cars and airplanes. And now, the six thousand-dollar paint job was down to thirty five hundred.

The next time I saw Ben's airplane it was a patriotic red, while, and blue, and now looked as new outside as it was inside. Now Ben was at peace with himself and with the world. And from that time on, every weekend he and his wife would fly somewhere and begin to enjoy their new 'magic carpet.' Sometimes it would be a trip and sometimes just a short jaunt for lunch.

Every time I ran into Ben during the next several months, he couldn't stop thanking me for helping him to get back into the air. He was the most appreciative person I've ever helped. Every time he'd see me, his smile would beam from ear to ear.

Ben is really a special friend.

•

Once Ben and his wife had been flying for some months, their trips became longer and longer, until finally they decided to visit their children and grandchildren. The next thing I knew, he and his wife were flying from El Monte, California to Spokane, or to San Antonio.

Time passed, and a year later, Ben bought a third partnership in a twin engine Beechcraft Baron. I think he wanted the use of a high performance twin to get to Spokane or San Antonio in a lot less time than he could in his Piper single. Especially on instrument flight plans, with minimum enroute altitudes at ten thousand and above, it just took too long to climb to cruise in his Arrow. At those altitudes, it was all the Arrow could do to just mush along.

With both a light twin or single engine airplane at his disposal, Ben settled easily into the role of weekend pilot. And it was during his weekend flying that he met up with a group of other weekend pilots who flew into Baja California, two or three times each year. The group was headed by an insurance broker who also published a quarterly newsletter for Baja pilots, called: 'Flight Log,' The newsletter was a six pager that focused on flying to Mexico to places where it was either to far to drive for a weekend, or where there were no roads. Places like Muleje, Punta Chivato, and Loretto.

Ben and his wife soon began traveling with the Flight Log group several times a year down to some remote locations of Baja California. When they returned after a long weekend, all Ben could do was talk excitedly about how wonderful it had been to just lay around in the sun and recuperate from the big city.

Ben and his wife had just returned from a trip to a place called Punta Chivato the previous weekend. As Ben described the trip, it was evident that Punta Chivato was a place so isolated you could only fly there by small airplane. Iit was halfway down the eastern coast of the Baja Peninsula, and there wasn't much there, except for a hotel built on a promontory some forty feet above the beach, with a landing strip next to it built out of gravel.

Ben couldn't stop raving about the place, and for weeks afterwards, he was still excited about it. He really wanted me to join up with the group for their next Mexican fly-in.

'The whole weekend, ya get to do whatever ya want. If ya want to just sit around and drink and get a tan, that's fine,' he said grinning. 'Actually,' he said in low tones, 'We divided up our time, between drinking, swimming, walking on the beach, and sun bathing. It's a great place just to relax and do nothing for two or three days. Ya could even go deep sea fishing, if ya wanted to.'

Over the course of the next year, I listened often to his Mexican adventures. I have to admit that lounging around, resting, and drinking sure sounded great. And after listening to Ben recount his flying weekends, I got the feeling that Barbara and I were really missing out on some good times. I decided that we too should take advantage of our airplane and join them on their next foray. A long weekend getaway seemed to be just the ticket — to be together with nothing to do but eat, sleep, drink, and sun.

The next excursion to Mexico was to be on Memorial Day weekend. Their itinerary was to fly from Southern California on a Thursday afternoon. They would then spend that night at the airport at Imperial, California, in a hotel close to the border of Mexicali. The next morning they planned to cross the border and clear Mexican Customs.

The next day, Friday, we would down to Punta Chivato from Mexicali, four hundred nautical miles non-stop. And after looking at a map, it was obvious that there wasn't any place enroute to land in an emergency or for fuel. The only place to land, once you took off, was at your destination of Punta Chivato.

Saturday morning, after spending Friday afternoon and evening relaxing at the hotel, we would fly from Punta Chivato to Loretto, seventy miles further south. It was said that Lorretto was the only place to shop for jewelry and leather goods in all of Baja. And then on Saturday evening back at the hotel, there was to be a big dinner for all of the Flight Loggers.'

Sunday was mostly a free day to just rest and drink and swim and sun.

And then on Monday, after breakfast, everyone would jump in their airplanes and head back to Mexicali, where they'd clear Mexican Customs, and then fly on to Calexico on the American side of the border, to clear U.S. Customs. And then from there the group would disperse on their separate ways.

So after learning the itinerary, Barbara and I decided that the whole thing sounded like fun and that we ought to go. Thus, I began preparing for the trip.

First, I had to join the Flight Log group, so I could get Mexican liability insurance from an American firm that was also recognized by the Mexican government. Next, once I joined, they mailed me a 'starter kit' for the trip, with all of the Customs forms so I could fill those out in advance, thus saving much processing time at the border.

Then I received an application form for the trip from Flight Log. I completed and mailed it along with a check for the first night's lodging. And there was also the matter of the administration and insurance fees to the Flight Log leader for his effort in coordinating the entire weekend.

I then had to get the 'Ship's Papers' in order for the Customs Departments on both sides of the border. Ben suggested putting the various licenses and forms in a small three ring binder. One with acetate pages and slip pockets for my pilot's license, passport, Mexican insurance policy, Airworthiness Certificate, Radio Station License, etc. For the Customs officers, would only have to open the binder to whatever page they wanted to inspect.

It was really a pain gathering all of the data, but once gathered, it would make future FAA ramp inspections a breeze. And when I was done, I put the binder in the map pocket behind the pilot's seat of my airplane.

Then I had to get a survival kit together. Not just the usual survival kit that was mostly for show and wouldn't keep a flea alive in an emergency. What I needed was a serious 'keep people alive for two weeks in the desert' kind of survival kit, one that contains two gallons of water and food for people for two weeks. A survival kit that costs a lot more than ten dollars.

Before this trip, the largest survival kit I ever needed for the kind of flying I did consisted of a quart of water, a bottle of scotch, and a Mastercard. Most of my flying was over civilized areas near interstates, from VOR to VOR, with an occasional mountain crossing.

The survival kit I would put together was a serious one, especially when I looked at the maps. Barbara and I would be flying over wilderness areas from Mexicali all the way down to Punta Chivato, some four hundred miles. There were no airports, if something went wrong, only mountainous terrain.

The aeronautical charts of Baja show that there was no landing places enroute for our trip. In the event that something did happen, and by some miracle, we were able to put down in the wilderness of the mountains without killing ourselves, then we would need what amounted to first aid medical supplies and camping gear to sustain us for at least two weeks. It would take us at least that long to hike to civilization. And in Mexico, civilization means the nearest village; and from Mexicali, the nearest village to the south, was San Felipe, some one hundred and ten nautical miles distant. From San Felipe, the next village to the south was another hundred miles.

Needless to say, the survival kit needed a tent, a stove, food, water, sleeping bags, signaling devices, and my King 99 two-way hand-held aviation radio.

An off airport landing in Baja wasn't the same thing as going down somewhere in America, since, in America, the Emergency Locator Transmitter or ELT that each airplane carries is automatically triggered when the airplane crashes. It can also be triggered manually if the airplane doesn't crash, but is forced down.

The ELT's signal is picked up overhead by the Search and Rescue Satellite or SARSAT, and the airplane's position is relayed to the National Search and Rescue Center, in Illinois. The Center then contacts the nearest Civil Air Patrol squadron, assigning them a Search and Rescue mission that would have them launching search airplanes in a matter of hours.

But south of the border, even if the ELT signal was picked up by the SARSAT over America, the Mexicans have no Civil Air Patrol, and if they did, I'm not sure they'd be eager to launch airplanes just to look for a couple of downed 'gringos.' If we went down between San Felipe and Loretto, we'd pretty much be on our own. So knowing that, I decided that what I needed was more than just a survival kit, I needed a collection of serious camping gear.

Years ago, when I read William Anderson's biography of the noted Alaskan bush pilot, Noel Wien, I was very impressed by something Wien said about going down in the bush in an airplane. 'When a pilot goes down in the bush, he becomes either a camper or dead; those are the only choices.'

And to survive in a desert or any climate, what one had to

do was to pretend that he was really going camping, and ensure that everything required to sustain life for at least two weeks was on board. Another consideration was comfort. Barbara really wasn't keen on flying in the first place, but she's always been a good sport and would go along if the destination was worth enduring the flight. Next to flying, on her list of least favorite things to do is camping. So when I thought about our survival gear, I knew it had to offer as much comfort as possible. Camping to Barbara meant staying in a 'three star hotel, instead of one with five stars.' It's not that she's spoiled, she just hates living outdoors. If all of the pioneers in Ohio felt like Barbara, they never would have 'headed west.'

So for comfort, I needed a roomy tent, two sleeping bags, two thick roll-up foam mattresses, water for two weeks, day packs for carrying food and water (if we survived the landing and could 'walk out), and clothing enough to keep us warm and dry. Additionally, I packed all of the signal devices I thought we would need.

I got all of the camping stuff together and loaded it in the airplane in one large tote bag, two-day packs, and a cardboard carton three-feet long and four-inches by four-inches for the tent. All of the survival gear weighed a total of fifty eight pounds, and took up half of the airplane's baggage compartment.

Now I don't know what Barbara expected of Punta Chivato, but according to Ben, at the end of May, it is hot and dry. The hotel sits up on a point of land that juts out into the Sea of Cortez; a kind of promontory. There are no other buildings. The nearest village is Muleje, fifteen miles to the south. The place is remote, isolated and completely informal.

While Barbara packed, I kept emphasizing to her that the place is totally informal. Tee shirts, jeans and bathing suits were what the guests wear during the day. As much as I tried to persuade her, it was to no avail, and in the end her luggage included two suitcases, a hanging clothes bag, a tote bag and a shoulder bag; just the bare essentials for a three day trip. In the land of jeans, tee shirts, and bathing suits, she packed as if we were never coming back.

All of my belongings were packed into two small tote bags.

Now my main concern was over grossing the airplane. The maximum gross takeoff weight for the airplane was twenty three hundred pounds. If it weighed more than that, it would perform in the hot Mexican air like a drunken Gooney Bird.

Besides too much weight, my other concern was a nagging preoccupation with the distance between Mexicali and Punta Chivato. It was four hundred miles of hostile territory and I continued to think about it and brood about it. Since I'd never flown my airplane on a long trip, I didn't have a good idea of its actual range, especially when it was over grossed. And now I was betting our lives that the thirty-year old Owner's Manual was telling the truth when it claimed that we could fly six hundred and seventy miles without refueling. Of course that was in ideal conditions at Standard Day temperatures in a brand new airplane.

The concern I had was with the weather, that it would not only be hot, but very humid, reducing the Musketeer's performance probably by twenty per cent. Even so, the reduction in range still gave us a safety margin of one hundred miles, which seemed plenty, even if we had headwinds both ways.

The entire route from San Felipe to Punta Chivato was over mountainous terrain with no roads or level places to land. The east coast of Baja from San Felipe to Punta Chivato consisted of shoreline that had no beach; just sheer dropoffs, where the mountains just sank into the water. It reminded me very much of Catalina Island west of Los Angeles where the sheer edge of the hills also sank straight down into the ocean, with not enough flat land at the water's edge for even a hang glider to land.

The more I brooded about the route, the more I focused on its dangers. I was beginning to 'psych' myself out. And pretty soon an ache started in the pit of my stomach; the one that told me that I ought to know the exact range of my airplane before betting our lives on the Owner's Manual. But, there wasn't time with my work schedule to take a long weekeend trip on a weekend to verify that our airplane could actually make it all the way to Punta Chivato.

From my flying experience, when that ache begins it's really my 'better sense' warning me that I've just traded away a margin of safety for an undue risk. The distance wouldn't have been a concern to me, if there was an airport halfway between San Felipe and Punta Chivato. But such was not the case, and my airplane was twenty nine years old, and for all the flying time I had in her, she'd never come close to the published cruise airspeed of the Owners Manual.

I calculated the fuel consumption and range several times, and each time the answer was that there was just enough fuel to make Punta Chivato in a no-wind situation. There might be as much as fifteen to twenty minutes of reserve fuel still in the tanks when we landed on the hotel's airstrip. Not much of a safety margin.

The stakes were too high and I began to feel like John Wayne in the 'High and the Mighty' when he realized that the fuel calculations didn't square with the book figures, and they might not have enough fuel to make Oakland from Hawaii with one engine out and fuel leaking over the wing from one of the airplane's ruptured fuel cells. He knew there was no way of knowing how much their real fuel consumption was with the added drag from a displaced and feathered engine, as well as how much the three remaining engines would consume, just to keep the airplane in the sky.

Only our trip wasn't a movie–this was real. This was

Barbara and me in our modest family airplane. The more I focused on the 'what if's,' the more I worried, and after agonizing over the Musketeer's range for a week, I just suddenly stopped thinking about it altogether; otherwise I'd scare myself into not going, when there was every reason to believe that the trip was possible.

Ben wasn't much of a confidence boost for me. He kept telling me about other low performance airplanes that flew down there regularly, like Cessna 172s that only had a fuel quantity of forty three gallons. Now normally that kind of talk would have placated me, but as I thought about Cessna Skyhawks, I wondered how heavily loaded they were, and their normal cruise speed was twenty-two miles per hour faster than mine, because of their more aerodynamically shaped fuselage.

●

The day before the flight, I filed a VFR Flight Plan for the next day, and not only did I file it, but I filed it for a 2 p.m. take-off time, after telling Barbara that I wanted to be airborne by Noon. I figured that if she procrastinated at all about her packing, then the two hour margin would be enough. But what I didn't count on was how much she really hated flying, and although she disguised her fear perfectly, the day of the trip slowed her actions down to super slow motion.

The day of the trip found us running around doing last minute errands of no importance. I knew that it was her way of prolonging her life–that unconsciously she thought that if she got in our airplane, she would die. So I played along and didn't say anything, as the time ticked away. And by the time we actually got away from the house and down to the airport, four miles away, it was almost 2 p.m.

I scurried about and performed the world's fastest preflight inspection, as the time pressure began to build, and as I thought about the flight plan about to expire in the FAA computer database. The thought of having to refile was now an additional aggravation.

Finally the time had run out; the extra two hours of pad that I'd given myself was history. I was becoming very nervous, inside. Outside, I acted as calmly as I could, so Barbara wouldn't feel any more anxious than she already did.

After I finished preflighting, I loaded our luggage on top of the survival gear already aboard. It didn't fit. I couldn't believe it. I unloaded everything from the baggage compartment, and even removed the rear bench seat. I put the rear bench seat into the truck of our car. I reloaded the airplane. There was still not enough room for everything. Something had to be left behind.

'Discretion being the better part of valor.' I left behind all fifty eight pounds of the survival gear. Out came the tent, sleeping bags, stove, food, and mess equipment; everything to sustain life for a stay in the wilderness. Even the signal devices were left behind. Only a plastic one gallon jug of drinking water remained, enough to keep us alive in the desert for four days.

Suddenly I knew how Amelia Erhart must have felt as she preflighted her airplane prior to crossing the Pacific for Howland Island, and had to leave behind the trailing wire antenna of her emergency low frequency radio because of weight considerations. Knowing full well, without it, she couldn't talk to any of the ships below that could save her.

After I removed and packed all of the survival gear safely into the back seat of our car, I then loaded our luggage into the airplane, all the while thinking about the four hundred nautical miles of desert and mountains we would be flying over. I wondered if Amelia got an ache in the pit of her stomach as she ran out of gas over the Pacific?

Taxiing to the runnup area was a hurried affair, even though I'd pretty much given up on filing a flight plan. The flight today would just be from Torrance to Imperial, mostly over the interstate until we got down to Oceanside and turned inland towards Julian that sat in the mountains. What ever margin of safety was lost by flying without a flight plan over civilization on the first leg of our journey was nothing compared to what we were giving up by flying south of the border without survival gear.

•

The day's weather was forecast to be 'severe clear' along the entire route, but as we departed Torrance Airport climbing to altitude over Los Angeles harbor, that forecast was not what we saw down the coast through the windshield. The beginning of this trip was very reminiscient of the beginning my solo cross country flight to Phoenix, as a student pilot, when the weather was also forecast to be clear and then wasn't.

The ache in my stomach suddenly changed from being a dull ache to a full fledged pain, as we turned southeasterly along the coast. There were a few low stratus clouds forming over the harbor, which was the beginning of fog. That wasn't supposed to happen.

And then a few minutes later, as we reached our cruise altitutde of seven thousand five hundred and leveled off, we were now seeing the beginnings of an immense fog bank forming offshore. From our vantage point it looked like the clouds below were a solid layer from the California coast all the way to Hawaii.

Something was happening and the weather was changing. It wasn't as forecast just hours ago and I didn't like it. I started to get a 'sinking feeling' to go with my stomach pain — an indication that my brain was telling me that the situation was dynamic and that I should forget my plans and call it a day.

I called Flight Service on the radio and asked about the changing weather. They took a long time to answer, but finally, a very 'laid back' voice replied that the weather was indeed changing and was not at all as previously predicted.

He said that the latest weather forecast concluded that a series of storm fronts of major proportions were approaching California from the east — a direction not normally associated with bad weather. But, it seems that there was a gigantic low pressure area over El Peso that inter-twined with the jet stream to funnel storm after storm towards California from the general direction of the Texas Gulf.

Usually, for days in advance, you can see storms coming towards California from the Pacific, but now for the next several days, the storms would be coming from the east, out of Arizona.

The briefer explained that, at the moment, there was a large thunderstorm developing over Julian, and that it extended southward along the crest of the mountains all the way to the Mexican border.

As Barbara and I listened to the voice in our headphones we exchanged glances. As he described it's size, we looked towards the mountains at our eleven o'clock position and could see the entire sky taken up with the largest cumulo-nimbus cell I'd ever seen in twenty years of flying.

The voice in our headphones droned on about the fact that it was already raining heavily east of Julian, in fact, it was raining all the back to Phoenix. And that a second large thunderstorm was already moving west from the Phoenix area, right behind the one we were looking at. Additionally, there were at least two more storms now forecast to hit California in the next several days.

'Right now,' the voice said, 'There's a rather large cell over Julian, and the rest of the series of storms are right behind it.'

My flight plan route was to fly over Julian at eight thousand five hundred, clearing the mountain peak by twenty five hundred, but as we neared Oceanside and turned eastward climbing to altitude, the thunderstorm was building into what seeme like the Stratosphere; maybe as high as sixty thousand feet.

The ache in my stomach was now a sharp pain like the kind you get with cramps. When the weather report was over, I thanked Flight Service and thought about Amerlia and John Wayne–things were not going well for us, at all. I had a very bad feeling about our situation.

I called San Diego Approach and asked them if they were painting any storm cells over Julian on their radar.

'Yessir!' came the cheerful reply. 'A great big one!' a husky feminine voice answered.

I hadn't expected a woman to answer, and for a moment the thought that a woman could help us caught me off guard, but then calmed my nerves a little. It must've been her relaxed tone that soothed me. For an instant, I tried to picture her in my mind so I wouldn't have to think about the thunderstorm in front of us. Was she a robust twenty seven-year old blond, or, maybe a brunette in her early thirties?' The whiskey voice sounded like it was 'filtered through a bedroom.'

'It's the biggest return I've ever seen!' she continued, sounding proud of her discovery.

I considered why her voice was so calm; she was sitting in front a very large radar screen, in a comfortably air conditioned room with subdued lighting. And she was not facing a thunderstorm of immense proportions, and if something happened to us, and we fell out of the sky to our deaths, she wouldn't feel a thing. No matter what happened, she was going to have a nice day. She wasn't responsible for our plight. Why shouldn't she sound relaxed?

Barbara and I could hear conversations between the controller and a Delta DC-9 traveling near Julian. The airliner was immediately vectored away from the cell. And then several moments later, a Cessna 210 fifteen miles ahead of us, following pretty much our routing, was also vectored away from the storm.

I looked at Barbara and she looked at me.

'Honey,' I began. 'Would you mind very much if I landed somewhere until I can figure out just how bad this weather thing is, and if it's safe to continue to Imperial today?'

'Not at all,' she smiled. 'I'm in no hurry to die.'

I could hear her sigh of relief over the intercom, as we stared ahead at the gigantic billowing clouds of the storm stretching along the tops of the mountains from wing tip to wing tip. The sight itself was mesmerizing, as if the clouds seemed alive as they billowed and seethed rolling about and shooting upwards.

'Anytime a DC-9 diverts from something, it's gotta be a hellava lot bigger than anything we could handle in this airplane and survive,' I said flatly into the intercom, thinking out loud for Barbara's benefit.

I had less than a minute to formulate a plan as we approached Julian, five miles ahead. Below us, the foothills were rising up underneath us at several hundreds of feet each second. And If we continued on our present course it would soon be too late to turn back and survive.

Suddenly a stabbing pain pierced my gut like a knife, and I had to stifle a gasp. For a moment I considered taking a vector around the storm, but down drafts in the area of a storm could easily force us downward faster than we could climb, and my Musketeer's performance was lackluster at best and no match for that.

The terrain was rising up so quickly beneath us now our

clearance above the granite was just a few hundred feet, and if we encountered a very swift down draft our lives would surely end before we could turn around and escape. Even the 210 with its high performance engine realized that it was time to give Julian a wide berth.

The choice was clear. We had to do a 'one eighty' and land somewhere civilized where we could relax and consider our options. The only place that came to mind from our present position was Palomar Airport at Carlsbad, some forty miles behind us.

Having been in and out of there over the years, I knew that Palomar was an upscale, controlled airport, used mostly by corporate jets and light twin commuter airplanes. It was nestled on a plateau with nearby hotels and restaurants. Because of the corporate activity, the airport even had an Executive Terminal on the field, where the corporate pilots hung out. The ramp in front of the place was always full of King Airs, Lear and Citation jets. And its pilot's lounge was furnished like the foyer of a hotel.

I knew for starters that Barbara would be impressed with the place since it wasn't the usual ratty-looking FBO. I had a good feeling about Palomar, so I keyed the mike on my headset and called San Diego Approach, as we banked to the left in a wide 'one eighty.'

'This is 93 Zulu. Can you give me a pretty good vector to Palomar Airport from my present position?'

'I can do better than that!' the female voice said, giggling. 'I'll give you a GREAT VECTOR to Palomar!'

And so, in the face of the overwhelming thunderstorm, we turned around and began a descent to Palomar, some forty miles to the west. And as we descended out of our cruise altitude I thought about the fact that if we did get weathered out of Mexico, at least we'd be close enough to visit our long time friends in Escondido.

As we backtracked, Palomar wasn't so easy to find below the fast moving cloud deck that was swiftly sliding onshore. The visibility was now forecast as scattered, but I could see that it was fast becoming broken, as layers of stratus overlaid and merged with those already onshore.

I dialed Palomar's designator into the LORAN in front of me, 'CRQ,' then pushed the 'TO' button once and the 'FROM' button twice. The display instantly appeared for Palomar Airport. I turned the large knob five clicks clockwise and the window displayed the 'Miles To Go, Ground Speed, and the Bearing ' to Palomar.

Soon the 'miles to go' were counting down as we descended through the developing overcast. When I called Palomar Tower they responded instantly, as San Diego had already alerted them of our intentions, after giving us a final vector and wishing us good luck.

Now beneath the overcast and on a seven-mile final approach, my spirits lifted and the ache in my stomach subsided And as we let down over low rolling hills a half-mile from the runway threshold, it occurred to me that the weekend might turn out completely different than we planned.

After landing, I parked in the transient space in front of Western Flite, the FBO for corporate flyers. It was an impressive facility and I hoped that its upscale decor would boost Barbara's spirits.

When we parked, ours was the only single engine airplane on their ramp; all the others were light twins or corporate jets. We slowly walked to the entrance of the FBO after securing the Musketeer. I followed Barbara through the large glass doors of the front entrance, hoping that it was a place where we could relax and consider our next move after checking out the weather.

A young girl in her late teens was behind the long counter

that took up one corner of the large room. The counter acted as a low profile office area. Behind the girl on a large desk return was a computer work station and FAX machine. The room itself was twenty by forty feet and decorated like a living room with five couches, four overstuffed wing backed chairs, and several low cocktail tables. It was used mainly as a pilot's lounge, and along one wall was a credenza where a silver coffee service and several silver trays were piled high with pastries. The place looked like a pilot's lounge; I could get used to.

As we stood there by the counter, I felt as if we were really 'somebody.' Of course, most of Western Flite's clientele were corporate airplanes or airplanes of high rollers, but they treated everyone the same

A hallway connected the lounge area with a side entrance that led to a parking lot, and located off the hallway were the rest rooms.

Barbara left me to use the restroom, while I poured myself a cup of coffee, grabbed a pastry and walked down the hallway to the Flight Planning Room to talk to Flight Service.

The room was complete with a direct phone line to San Diego Flight Service. A shelf along one wall held more published source directories for the flyer than I knew even existed, enough to plan a flight to anywhere in the world. A computer workstation for DUATS was located on a desk at one end of the room. The computer accessed the FAA's weather briefing database. With the press of a key, it displayed the terminal weather at any reporting airport of any country in the world, including NOTAMS and PIREPs. It was a real self briefing station, the kind the FAA wanted to establish everywhere, but never did. The common wall of the room with the hallway was all glass; and both VFR and IFR planning charts of the U.S. were mounted on it.

It was obvious that a pilot could walk into the room without even a pencil, and walk out with a FAX'd weather map of his newly planned flight routing. He could also receive the terminal forecasts for as many airports as he desired along his route of flight, and a copy of his filed flight plan with the FAA. And the entire service was free to the flying public.

•

When I got Flight Service on the phone, I told them what had happened to us weatherwise, and asked what the prognosis was for our route of flight. I listened for a long time without saying anything, just offering an occasional grunt now and then, so the briefer would know I was still there. The voice began a twenty minute soliloquy that became the most detailed weather briefing I'd ever received. I think the briefer was afraid that the news of bad weather wouldn't impress me enough to sit it out instead of 'testing' the forecast.

For me, the story was grim enough and he didn't need to make it sound any worse. The forecast was for a series of storms coming from the east that were actually being generated by a large low pressure area over El Peso that was picking up moisture from Baja and the Gulf. It was then generating a series of thunderstorms from Houston to Albuquerque to Phoenix to San Diego to halfway down Baja. At the moment there was moderate rain showers all through the Imperial Valley, where we were supposed to spend the night with the others of Flight Log.

The next day's forecast wasn't anymore promising; but it was still too early to tell for sure. The briefer asked me to call in the morning after six thirty for a better forecast for tomorrow–but he thought it would be a smart idea just to sit tight for the rest of today, now that we were on the ground.

By the time Barbara returned from the rest room, I'd discovered that the best hotel within twenty miles of the airport was just down the street. The receptionist wouldn't let me call a cab, since one of their mechanics was leaving in a few minutes on an errand and could easily drop us off at the hotel.

While we were waiting for the mechanic, I phoned the hotel to see if there were any rooms available on this Thursday, two days prior to the long Memorial Day weekend. And as luck would have it, there was one room still available. I took it sight unseen, as I've learned when traveling with my wife, that 'the best there is, is just about good enough,' when plans change while enroute on a trip.

The mechanic picked us up in a van and five minutes later, just down the road a half-mile from the FBO, just abeam the approach end of Runway 24, was the Olympian Resort and Spa. The hotel was a former spa now reconfigured to a combination golf and tennis school. The school was famous for its three and four day packages for the serious player. The facility was complete with classrooms, closed circuit TV, and video tape players, so that each private lesson could be taped and the player's swing analyzed and critiqued.

Our room, as it turned out, overlooked the golf driving range on the backside of the hotel. The driving range had tee positions for seventy-five; complete with night lighting, and a pro shop with two resident pros. The tennis school also contained the similar facilities with twelve tennis courts and a pro shop with two pros. Both schools were really big operations.

From the balcony of our second floor room overlooking the driving range, we could watch the 'duffers' trying to improve.

After we got to our room and unpacked, I phoned a nearby car rental agency, and within a half-hour a rental car was delivered to the hotel. With rental car keys in my pocket I didn't feel stranded anymore, and looked forward to an interesting weekend if the weather didn't clear for our flight to Mexico.

That night we had dinner in the hotel's dining room and thought about calling our long time friends in Escondido, just a half-hour away. With the weather changing we weren't sure just how long we'd be staying at the hotel, but since we were so close, we decided to give our friends a call, and maybe drive over for a visit.

So by the end of the first day of our Mexican weekend we found ourselves only one hundred miles south of Torrance Airport, with the weather overcast and drizzling and the prospects for getting to Mexico the next morning, in doubt.

•

The next morning I was exhuasted from visiting our friends until midnight. It had been a long time since we'd seen them and needed to rekindle a friendship too long ignored. As tired as I was I dragged myself out of bed by 6:30 a.m. It was now time to get the weather forecast from Flight Service in San Diego, that would dictate where we would spend the remainder of the long weekend.

To check on the local weather I opened the shade and looked out the window. It was raining and overcast. The skies hadn't changed since the evening before, when we drove to Escondido. So, I held my breath while the briefer punched the keyboard of his computer for the latest forecast for our route.

'It doesn't look good,' he began. That line was right out of the movie 'The High and the Mighty.' And as he talked, I wondered if our flight would have a similar fate.

'First of all,' he continued, 'The satellite picture shows overcast skies from Phoenix all the way to Hawaii. And down Baja way, the low over Houston is sucking sub-tropical moisture from the Pacific, west of Baja, that obscures the entire peninsula.

Only the tip of Cabo San Lucas is clear of the overcast.

'How high's the overcast down there?' I asked, thinking out loud.

'Dunno. It could be real low or four or five thousand. We don't get reports from down there, so there's really no way of knowing.'

'Well, how does it look from Palomar to Calexico, in a couple of hours, say around 9 o'clock this morning?'

'I'll tell ya the truth,' he continued. 'If it was my wife and me that were heading south for a long weekend, I'd forget it. This isn't the weekend for an unforgetable vacation. If it was me, I'd have a wonderful time somewhere else. Even if you're instrument rated, I'd still scrub the trip.'

'What do ya mean? I thought the flying down there was always VFR.'

'I know that's what they say, but what about the return trip on Monday. Not only would you be real marginal, you'd probably be in instrument conditions for sure. 'Consider this,' he said with in a serious tone. 'If the satellite can't see the ground, from it's position in the sky, then there's sure no way to tell how thick and low the overcast is. And I'm talking about an overcast that goes from Mexicali all the way down to La Paz. And these series of storms aren't due to clear out of that area for another four days. I mean, this is one immense and continuous 'hummer!''

'I got the picture,' I said, the trip beginning to vanish before my eyes. 'But can't ya assume that the overcast is high enough to fly under?'

'You're taking your wife on a trip, where you're assuming the weather's just good enough to sneak under?

'I get your point,' I said, resigned, and thanked him for his candor.

Barbara woke up just after I finished talking to the briefer.

'Well?' she asked, sitting up on her elbows. 'Do I get up, or turn over and get another two hours sleep?'

'Go back to sleep. The weather's awful for another four days. I'll call Ben at the hotel in Imperial before he leaves, and tell him we're not able to get out of Palomar, let alone to Mexicali.'

And so, I phoned Ben who was just leaving for breakfast, and told him about the forecast and that I didn't want to scare Barbara to death by flying through rain and turbulence. 'It's been raining here all night in Carlsbad, and the forecast is for more without a break for the next four days,' I explained.

'Listen Bill, Ben pleaded. 'Pay no attention to the rain. It's been raining here at Imperial since yesterday, but I think if ya come down to San Diego along the coast, and then turn east and following the border to Calexico, ya shouldn't run into any thunderstorms to speak of.'

'After yesterday over Julian, all I can say is not to count on us. I don't want to scare Barbara. I want her to like flying, if that's possible. A flight through bad weather, for me, just isn't worth it.

'Well, really try, okay?' he said. 'We're all going to be disappointed if ya don't make it.'

'I know, but I'm also worried about coming back on Monday. This isn't just one storm–it's a whole string of 'em, one after another. Even if we get down there, I'm not sure I could make it back on Monday.

There was silence on the other end.

'If we don't make it this time, there'll be other times, Ben,' I said, trying to ease out of it as gently as I could. I knew that if I didn't feel good about flying to a strange place in bad weather, I sure wasn't going to fly there with my wife.

Day two of our trip to Mexico was again spent visiting with our friends in Escondido. And as I was sipping a glass of twelve year old scotch, the thought occurred to me, that sometimes things are predestined and no amount of planning, scheming, or maneuvering can change our destiny.

•

When we awoke in the hotel on Day three of the trip, Saturday, it looked just like Friday. It was still overcast, but the wind and drizzle had stopped. Barbara and I got up just in time to pick up our friends and drive them to a restaurant in the 'old town' part of Carlsbad. After lunch, we passed the afternoon from the patio of one of the quaint old taverns on the boulevard. And by mid-afternoon we window shopped our way through the numerous boutiques. And ended the day in Escondido with a backyard barbecue.

As we were eating lunch, the overcast began to break up enough to let streams of sunshine brighten the afternoon. Later, in our friend's backyard in the cool of the evening, standing by his barbecue, watching steaks fry and sipping another tall glass of single malt scotch, I wondered if Ben and his wife were having as good a time in Punta Chivato, as Barb and I were having in Escondido.

So, by the end of Day three, Barbara and I had done exactly what we intended to do in Mexico: relax, rest, and enjoy a weekend of simple pleasures in the company of friends. But instead of recreating in Mexico, our venue was Carlsbad, California.

That night as we ended our visit, we said our final goodbys and returned to the hotel. It was a good feeling to rekindle a long time friendship.

•

The next day, Sunday, we checked out and drove to the airport. The skies were still overcast and a gusty wind had come up, as if another front was coming through. When we got to the FBO I called Flight Service and was told that the entire California coast was below VFR minimums and would stay that way for several more days. So, undaunted, Barbara and I got back in the rental car and drove home, leaving the airplane at Palomar Airport; the final insult to a VFR pilot.

•

It was on the following Sunday, a week later, when my instrument flight instructor, Guy Gillmore and I finally got back to Palomar to rescue the Musketeer. Barbara drove us down so we could fly the airplane back to Torrance. Even though it was a week after the holiday weekend, the weather was still IFR along the southern California coast. For once, Flight Service weather briefers really knew what they were talking about.

Guy and I flew the airplane back to Torrance on an IFR flight plan and discovered that, as we climbed to our cruise altitude of six thousand, the tops of the cloud deck that comprised the overcast was at five thousand feet. But as we looked to the west there were no breaks as far as we could see, and maybe it extended all the way to Hawaii, after all. But at six thousand the sun was out and shining brightly.

The weather didn't improve much for another week — two weeks after our trip was finally over.

•

Back at work, Ben told me that the overcast was fairly high going down to Punta Chivato on Friday, but coming back on Monday, it was almost marginal VFR with just a mile visibility all the way northward to Mexicali. And by the time they cleared Customs at Calexico, on their way back to El Monte, the poor visibility, high head winds, and rain forced them to file an instrument flight plan and return on the gauges.

Ben told me that on the return trip the head winds were so strong that he was beginning to worry about his fuel consumption. Needless to say, after I heard that, I knew that I'd made the right decision not to attempt the trip beyond Carlsbad.

•

As I look back, Barbara and I had a really wonderful time even though we never did get to Mexico.

I don't think Ben ever forgave us for not braving the largest series of storms to hit California in years. But, somehow, even if we had been able to get to Punta Chivato, I don't think we would have had quite as much fun, and besides, in Carlsbad we were able to drink the water!

Chapter Eight
Hangar Flying Without a Parachute

Flying Adventures at Ground Level

For Sunday Pilots, 'hangar flying' may be defined as the earth-bound activity of flyers talking about flying to other flyers. Historically, hangar flying originally took place in hangars, when bad weather forced pilots to sit around talking about airplanes and flying to pass the time. However, present day hangar flying can take place anywhere two or more flyers get together to share their flying experiences.

The following characters represent some of the best hangar flyers I've known over the years. I hope you enjoy them as much as I have. To several, I owe a special thanks for motivating me to continue flying, when I'd really lost the spirit and wanted to quit. I consider them my special friends. Some were merely flying acquaintances who offered me comic relief in a sometimes-dreary world. But, all of them are my 'band of brothers,' whose memory I will cherish, always.

H. Peter Stoterau, Esquire

As he sat on the corner of my desk, H. Peter Stoterau, Esquire, or 'Pete,' for short, motioned with his right hand to illustrate the attitude of the Cessna 210 he was telling me about, as it descended on final approach to Runway 8 at the Burbank Airport. His friend Hal owned the airplane and was at the controls, after taking Pete for a ride. The airplane was descending easily down the glide slope as Pete's body squirmed in the right seat. He'd never flown a 210 before and was unfamiliar with its performance, spending most of his leisure time flying Cessna 182s.

Pete sat watching Hal set up a good instrument approach from the outer marker, located at the departure end of Runway 16 Right of the Van Nuys Airport, seven miles west of Burbank. They crossed the marker at three thousand feet, as Hal waited impatiently for the glide slope needle to begin to move. Pete glanced through the windshield at Runway 8, now about five miles ahead. It was fairly easy to see, if you knew where to look. Since the localizer pretty much was aligned geographically with Sherman Way Boulevard, all one had to do was to follow the street until it disappeared — dead ending into Runway 8.

Pete watched the view of Burbank Airport grow in size, as the 210 approached the runway at over one hundred and fifty feet per second. It was summer and the air was hot in the San Fernando Valley. As they passed through Decision Altitude, the beeping tones of the middle marker were just fading as Pete looked over at Hal, who was sitting there transfixed, making no attempt to arrest the descent angle by rotating the airplane's nose to level. With the gear down and half flaps, they were coming down out of the sky with the gliding characteristics of an anvil.

"Hal?' Pete queried, as if to arouse the pilot from his stupor.

Hal didn't respond. Instead, he sat frozen at the controls. At their present speed, if the airplane's attitude wasn't arrested within seconds, it would crash just beyond the 'touchdown zone' markings painted on the runway, and the Cessna 210 would turn into 'balled aluminum.'

I squirmed in my chair, as I visualized the scene. As he spoke, Pete brought his other hand into play, while I sat smiling and alternated my gaze from his eyes to his hands. Pete was a master story teller and it was hard not to get engrossed in whatever he was explaining.

'I didn't know what to think,' he said frowning. 'Our airspeed was 90 knots, and my buddy, Hal, made no effort to bring the control wheel back. I'd never flown with anybody that 'froze up' at the controls, but when I knew we couldn't wait any longer and he still didn't respond, I yelled 'I got it' at him and took the controls and simultaneously pulled the control wheel back and pushed the throttle in to add a little power.'

At this point, Pete's right hand, pointing down at a thirty-degree angle, suddenly rotated to the horizontal, while his other hand, palm down, illustrated the runway.

'I immediately leveled the nose and then after a few more seconds, eased the power back to about fifteen inches and then just held that attitude. We took up most of the runway before I could get the thing on the ground and stopped. But as soon as I taxied off the runway, I looked over at Hal, and told him 'to take it.'

'What happened to him?' I asked.

'I don't know,' Pete replied. 'He thanked me later for saving us. I think he just froze up mentally. I think he was too embarrassed to tell me the reason.'

'Do ya think he blanked out, like a mini-stroke?'

'I never did ask. You know how it is with a close friend. Some things you just can't ask about.'

•

When it came to story telling and especially stories about flying, Pete could always find just the right words to describe whatever the situation warranted. He was really a master at it. But then, he was a lawyer and words were his stock in trade.

It was almost noon, and Pete had stopped into my office for twenty minutes to do some 'hangar flying' just before lunch, since he and I were the only pilots in the entire law firm of McCutchen, Black, Verleger and Shea. The firm's offices were located in the mid-Wilshire district of Los Angeles, three miles west of the downtown area.

I joined the firm the year before as the Administrator, after a fifteen year career in the Engineering Department of Douglas Aircraft Company of Long Beach. I wasn't a lawyer, but as an engineering manager at Douglas, McCutchen hired me to manage the daily operations of their general staff–everything that didn't involve the practice of law.

At the time, McCutchen was a middle-sized law firm with seventy seven lawyers: twenty seven partners; and a general staff of one hundred and fifty. The managing partner of the firm needed to be relieved of the daily operations so he could spend more of his time practicing law. And, although I didn't know much about law, I took the job because I knew something about managing people, labor unions, administrative operations, and the usual Human Resources tasks of personnel, benefits, and compensation.

My new office at McCutchen was large compared to the one I'd left at the Douglas Aircraft Company. It was located on the top floor of the building, and contained two windows that faced north. The office came with thick carpet, dark mahogany furniture, and a large high-backed leather arm chair. For me, it was quite a change from the 'early aerospace' decor of an airplane factory, with its metal furniture painted gray, and desks with linoleum tops, linoleum floors, and no windows.

I guess I'd been on the job at McCutchen for about a month, when the word got around that I was a pilot. The next thing I knew, H. Peter Stoterau, one of the twenty seven partners, walked into my office one morning and introduced himself. And before I could even say, 'It's a pleasure to meet you,' he sat down on the corner of my desk and started talking about airplanes. It seemed that he'd done a hitch in the Air Force as a pilot, and was desperate to find someone to 'hangar fly' with.

Needless to say, we became immediate friends.

Physically, Pete was a big man, standing six feet six inches tall and weighing in at over two hundred and thirty, and his deep resonant voice commanded instant attention. He was in his mid forties and had spent his early years at the firm in the Litigation Department, defending various Corporations. With his deep voice and large frame, he surely posed a dynamic 'presence' in court, something akin to Orsen Wells, in the movies.

That was long before he became a partner, and he told me once, that he really didn't care much for the life of a litigator, since it wasn't really his style to make everything into a contest. He much preferred what he was doing now, working in the Business Department, putting large business deals together for the 'movers and shakers' of the city.

Now his life consisted of business meetings over lunch, the golf course, or the boardroom; or conversations by the hour over the telephone. His forte was negotiation, and others of the firm remarked that he was very good at it; a 'natural' negotiator, the kind of salesman that could talk anybody into anything.

Since I'd been crazy about airplanes and had worked for an airframe manufacturer, I missed being around them. Now I was working downtown in the 'business world.' Before, when I worked at Douglas Aircraft, I was pushing paper, but at least I could always walk out to the flight ramp now and then, and see the airplanes that the company produced, and even talk to the Company Pilots or the Dispatchers. I was downtown, where there were no airplanes, and my only connection to aviation now, was the random sessions of 'hangar flying' with Pete.

It was a strange period of my life. I had stopped flying during the last year and a half while I attended graduate school at night getting a Master's Degree, and had neither the time nor the money to go flying. And since getting the degree and leaving Douglas, I hadn't really been motivated to take up flying again, until I met Pete.

Since my flight training in the early seventies, all of my flying had been in rented airplanes and I was tired of the expense. And it became hard for me to get motivated enough to stay current as a pilot–to fly frequently enough to completed three 'touch and goes' during the last ninety days. It doesn't sound like much, but when there's no fun in it, and it costs over a dollar a minute to do it, and there's no reason to do it other than to stay current so you're able to rent airplanes whenever you want — what's the point?

And sooner or later, the motivation to fly at all disappears.

Flying rented airplanes had taken on kind of a boredom for me, and although I was still crazy about flying, I couldn't bring myself to spend the money even for an hour's ride. I had a nagging preoccupation about flying, but I wasn't sure what to do about it. I really felt like something was missing from my life; but if the missing piece was the kind of flying I'd been doing, the kind that wasn't fun for me anymore, then it would have to stay gone.

Pete's energy and enthusiasm was contagious, and after a year of hangar flying with him, I got motivated enough to get checked out in Cessna 172s. It was a change for me, having spent my entire flying life in low wing Pipers. Now, in high wingers, I could even see the ground as I flew along — a welcome change. But, flying rented airplanes was still boring for me, no matter what the wing configuration was. It was boring mostly because I really had no destination or reason to fly anywhere; but I was hoping that somehow I could turn that part of my psyche around.

Pete was an interesting guy and knew some big and powerful people in the airline business; he had represented New Zealand Airlines in litigation several times. He also was on a first name basis with the upper management of Continental, American, and United Airlines. And every time he was in New York or Chicago or San Francisco on business, he'd stop in and say 'Hi' to his many 'flying buddies' of the airlines; where upon, if they had the time, they'd drop everything and take Pete flying for a couple of hours.

Pete's manner was infectious; you either immediately liked him or didn't. Although, like most lawyers, he could be overly assertive and egocentric and manipulative, he also had a certain 'Tom Sawyer' charm that made the difference. The other thing that I noticed about him, was that he didn't give the impression of being superior to the 'mere mortals' of society, as most of the other attorneys of the firm did. He'd just walk in and sit down on the corner of your desk, like he'd known you all his life.

Hangar flying, besides the sharing of lessons learned experiences, is also a sharing of flight itself. It makes up the mythology that the flyer needs to keep motivated and his 'batteries charged.' In the sharing of the experience, whether it's a personal flight experience or someone else's, the torch has to be refueled to continue to burn.What has always been interesting to me is that, it doesn't matter where you are or at what station in life, when two flyers get together, hangar flying happens.

I'm sure it's the same in other hobbies, when two people

with similar interests discover each other, and share; but for me, I owe Pete my gratitude and thanks for relighting my 'torch' and 'rekindling' my interest in aviation.

Joel Premsaalar

Joel Premsaalar was the perfect image of a 'Dutch Uncle.' Although he was Dutch by birth and looked the part, that's where the distinction ends, as his manner was never unflattering or direct to the point of 'telling it like it is.' His persona was quite the opposite, as his style was that of a 'silver fox' with the ladies, even though he was in his mid-sixties.

He had a portly build with chiseled features; all he needed to look the part of a Dutch Sea Captain from Central Casting was a Captain's cap, the kind with a large gold anchor over the bill. Actually, the image of the sea is partly true. He joined the U.S. Navy in the mid-thirties, as a teenager, and worked his way up the ladder from tending boilers to becoming one of the first 'enlisted pilots,' in a special WWII program.

In many ways, Joel reminded me of Tom Stansbury, since women just seemed to flock to him and 'moon' over him without the least encouragement on his part. I was never jealous, even when Barbara chatted with him once, at a friend's backyard barbecue. She only talked to him for a brief period, but years afterwards when his name was mentioned, she would smile, remembering their encounter.

•

Joel had flown every propeller airplane the U.S. Navy had during World War II. And afterwards, he transitioned to jet fighters during Korea. After Korea, he was stationed at China

Lake Naval Weapons Test Center, as a test pilot, until his retirement from the service. After he retired, he went to college and four years later, earned a degree in Aeronautical Engineering, where upon he was hired by Boeing in Seattle.

He told me one time that his fondest memories of flying were of the early years in the Navy, when he piloted the only airplane aboard a cruiser. It was a single pontooned low wing scouting seaplane that the ship would launch from a catapult off the fan tail. Since he was the only flyer aboard, he really felt special. When they launched his seaplane, every eye was on him. When he returned from a mission and would land near the ship, everyone would again turn out to watch. The crane would lower a cable over the side from a hoist, and ignominiously drag the airplane out of the sea and then lower it back onto the fan tail.

I learned a lot from hangar flying with Joel. He'd been everywhere and flown every kind of airplane. Not only had he flown everything, he'd escaped from more dangerous situations than one lifetime holds. He was also a master story teller and teacher, weaving a 'lesson learned' into the fabric of every scenario and explaining in great detail how his mind deduced 'the situation' and reasoned a plan of action. It was like being in a scientist's laboratory after he'd discovered a cure for something, and you got to read his notes documenting the procedural steps of his experiment.

But what made me really appreciate Joel's acquaintance was his intellectual unselfishness which I discovered after working for him as a Systems Engineer. His one shining quality as an engineering manager was that he wasn't at all selfish about information, and would share the latest technology news with everybody. He also wasn't stingy with accolades. We were working on an R&D project, where breakthroughs are coveted and only the boss takes the credit. But with Joel, it never worked that way, if you were on his team, whenever the opportunity arose, he would give public credit to those doing the work.

'After all!' he'd say. 'There aren't any jobs in the world to be envious of, after you've been a Naval Aviator. And I don't need any more accolades; I've had more than my share for one lifetime.'

•

As part of our hangar flying regime, Joel would quiz me about various aviation phenomenon, such as: 'when is the best time of day to refuel an airplane, to get the most BTUs for the money?' He'd ask the question, as if he didn't really know the answer, when of course he did. He made the quiz into a learning experience, in which the lesson hit you like an after thought. He was so smooth at it, I could see why he was such a hit with the ladies; he was like a Venus Flytrap–the ladies never saw him coming.

Although I knew Joel was a pilot, it wasn't until I'd been working for him that he told me that he'd bought a vintage Beechcraft Bonanza. Now, he'd owned a number of airplanes during his life, and this was his second Bonanza.

I'd owned my Pacer for several years before I was on Joel's staff, and during that time, Joel and I used to talk about airplane maintenance. It was always his contention that if a pilot helped with the annual inspection of his own airplane, then he'd gain a deeper appreciation for how complex even the simplest airplane was — once he discovered how every system worked.

For several years I'd been paying five hundred dollars to an FBO to have the Pacer 'annualled.' The FBO's motto was: 'It's five hundred to taxi into the hangar and say hello; and then parts or labor for fixing anything that needs it, is extra.' So when Joel talked to me about trying an owner-assisted annual that cost half as much as what the FBO charged, my ears perked up.

The owner-assisted annual inspection is one in which the owner does all of the work, while the mechanic supervises and 'talks' the owner through each task, directing him in the proper technique and methods.

'Well, where can ya find a mechanic that does owner-assisted annuals?'

'You mean you're finally getting around to wanting to learn how your airplane works?' he laughed.

'Yeah, I thought I'd take ya up on the idea, and even save a couple of bucks.'

'Well, you'll save a lot more than just a couple of bucks, and you're liable to learn a lot more than you think, about your airplane,' he smiled. 'Wasowski's the name. Walt Wasowski. He's got a hangar two down from mine over at Compton Airport. Fly your airplane over there this Saturday morning and look him up. I'll tell him to expect you. And you won't be sorry!'

And that's how I came to meet a mechanic named Walt Wasowski.

Walt Wasowski

Walt Wasowski was a natural 'hangar flyer,' having actually been a private pilot prior to World War II, in the late thirties. But he was an airline mechanic by trade, and it's my hunch that he really hadn't flown an airplane since Pearl Harbor. For the last thirty years, he'd worked for American Airlines, and as he neared retirement, they used him as a kind of 'trouble shooter,' sending him off to Dallas for a week, and then on to Kennedy for a month and then back home to Los Angeles.

For a Polish person, Walt had a gift of gab, reminiscent of the Irish, and it soon became apparent that he'd used his communications skills his whole life to get out of one scrape after another. When anyone asked him a question, he'd begin a monologue that would last for an hour. For example I asked him how he liked working for American Airlines. Instead of telling me that it was great or not so hot, he described his entire career with them. It took 25 minutes, and that's how I found out how he'd gotten started in fixing airplanes.

He'd been an airplane mechanic, self-taught, before World War II. After Pearl Harbor, he enlisted in the Army and they had him training aircraft mechanics on the job, somewhere back east, until the Army established formalized training schools. But, the Army couldn't spare him so that he could attend the school himself; they said he was needed where he was, and besides, he already knew more about fixing airplanes than the school could teach him. So, instead of sending him to school, the Army promoted Walt to Technical Sergeant, and within a year he was in charge of all the airplane mechanics at a B-25 Bomber base.

It was easy for me to see how Joel and Walt got along, since it seemed to me that they were cut from the same cloth. And I really never knew whether to believe everything that Walt told me. He was like Joel; a 'silver tongued devil.' Besides, whatever he said, Walt always reminded me of a Leprechaun, since he was just five foot three inches tall, with a distinguished portliness, and a head full of thick white hair that he kept trimmed close along the sides, like Army 'whitewalls.' The white hair gave him the appearance of a man in his fifties, rather than one in his seventies. And every time I thought he was pulling my leg with one of his stories, I'd check the story with Joel, and come to find out that Walt had really done whatever it was.

I met Walt one Saturday morning at Compton Airport. Joel told me that Walt would be performing an annual inspection on Joel's Bonanza, and that's where I could find him. So Saturday morning I flew my Pacer over to Compton, from Hawthorne Airport, a total distance of six miles or four minutes by air. I wasn't sure of the exact location of Joel's hangar, so I parked the airplane at the end of the northern most row of hangars and started walking until I saw a hangar with an unpainted Bonanza in it, and a fellow laying underneath its left wing.

'Excuse me. Is this Joel Premsaalar's hangar?'

'Ya, this is it,' a voice yelled back from under the airplane.

'Are you Walt Wasowski?'

'That's me.'

'Joel sent me. My name's Bill, and I'e got a Pacer that needs an annual. Joel suggeted I look ya up.'

'Where's your airplane?'

'It's down at the end of the hangar row. I didn't know where to park it'

'Well, I'll be right with ya in a couple of minutes. Go taxi your airplane down to my hangar. It's the one that's open with the machine shop in it.'

So I did as I was told, and about half an hour later, Walt came walking back from Joel's hangar, and introduced himself shaking my hand. He told me he could do my annual the next weekend, but for me to arrive early Saturday morning so we would have two full days to work on it. It would be completed by Sunday afternoon.

'Well how will I get back to my car at Hawthorne, if I have to leave it over night?'

'I'll drive ya over, I live in Hawthorne. It's on my way home.'

And with that, Walt and I began a four-year relationship that lasted until I sold the Pacer.

•

Walt was the kind of mechanic who knew what had to be done and would lay out a plan off the top of his head and begin opening up things or removing things in preparation. But he would forget to verbalize what the intended item of maintenance was. Or, he'd be quite the opposite–telling me in great detail what to take off, in what order, and then what to do for the next hour. After explaining in great detail, he'd then leave to take a nap for an hour before returning to check on how the work was going.

One of Walt's peculiarities was his cavalier attitude. I could never tell when he was serious. He had a habit of second guessing the FAA's reason for issuing Airworthiness Directive or ADs, or a Manufacturer's reason for issuing a Service Bulletin against a specific model airplane. Normally, these are issued because an unsafe condition exists that must be remedied prior to the next flight. Walt's idea of a 'safety of flight' item, and the FAA's, usually were world's apart.

'I remember the time that the Piper Pacer had an AD issued against its wing strut forks. This is the fork that connects the two wing struts to the fuselage. The FAA wanted the forks 'dye tested' every five years for corrosion and cracking. To me that seemed to be rather important, although no Short Wing Pipers had ever lost a wing due to the struts failing. But, after all, the struts were now over thirty years old on most of the Pacers still flying, and even though they were sealed and usually filled with mineral oil at the factory, to prevent internal corrosion, thirty years is thirty years.

'And when the logbook said it was time to test the struts, I fully expected Walt to get a dye kit and paint the struts to test for any cracks. When I mentioned the test to him during the

Annual that year, all he said was, 'That test doesn't show ya nothin.' I got a better test than that. Watch.'

Slowly and deliberately he walked over to the right wing strut and took his foot and kicked the strut fork as hard as he could, where the strut attaches to the fuselage. The fork, at the junction of the two wing struts and the bottom side of the fuselage didn't move, but the struts gave out with a soul wrenching 'Wowing' sound, as if the 'sudden indignity' of Walt's boot was felt deep within the airplane's soul.

After he did it, he looked over at me grinning. 'Ya see? Just as I thought. Nothin.' Those struts ain't ever coming loose. My kind of test is the only test ya need,' he laughed, walking over to the workbench where he kept the airplane's log books, and signed off the test in the airframe log as having been completed and successful.

From that time on I was never sure about those struts, and a year later I sold the airplane. Fortunately for me, the new owner completely 'reskinned' the airplane, installing new struts as a safety measure. I also wondered after that 'test' about how thorough an inspection Walt really accomplished on any airplane. When I sold the Pacer, not only did I breath a sigh of relief about the safety considerations of old airplanes, but I gave up Walt as well. Old airplanes are fun to fly, but to keep them maintained and safe is a serious matter.

•

Bob Root

Bob Root was a svelte six feet two inches tall and handsome. He had thick prematurely gray hair that was bordering on white. There was a certain 'boyish' quality about Bob that women were attracted to, like a kind of grown up 'Tom Sawyer.' It was a different allure than Joel's, although just as effective. Bob, always reminded me of Robert Redfern with gray hair; and he was always surprised when women would 'come on' to him while he just went about his business. He really was oblivious to his surroundings. And he always seemed preoccupied and unaware of the female stares.

Bob was just ready to take his checkride for the Private Pilot License when I gave him a ride in my Pacer. As it turned out, he'd bought a 1949 Piper Clipper, the model just previous to the Pacer. Except for the fact that the Clipper's engine was a little smaller and it had a control stick instead of a wheel, the two airplanes were pretty much identical. The Clipper was lighter and its tail wheel was six inches in diameter instead of ten.

We took off from Hawthorne Airport at lunch time one Wednesday, and after climbing to one thousand five hundred feet and aiming for the Los Angeles harbor, I turned the controls over to Bob. He said flying an airplane with a control wheel felt strange for him, as the Clipper has a stick; and I had to show him which hand to use for the wheel and which to use for the throttle. But, to be fair, he was flying it from the right seat, which for a student pilot used to the left seat made things feel different. But he was a fast learner and within a couple of minutes he got the hang of things. Soon, he was flying some turns to the right and left and experiencing how sensitive the rudder really was.

We flew for an hour over the coast, from the harbor down to Newport Beach and back. Bob seemed to be a 'natural flyer' and like me, had been crazy about airplanes since childhood.

During his life, he'd been an avid airplane model builder, and as a hobby during his early years as an Aeronautical Engineer, he'd gotten into Radio Controlled model airplanes, or 'RC Models,' as they are called. He told me one time that he even made the National Championship finals for 'RCs', held at Wittman Field in Oshkosh, the home of the Experimental Aviation Association. But now, he was flying real airplanes, owning a Clipper and had it in a hangar at Compton Airport.

When I heard that he was based at Compton, I immediately asked him if he knew Walt Wasowski and Joel Premsaalar.

'Ya must be kiddin,' he laughed. 'Joel introduced me to Walt, and Walt's now my mechanic.'

'Me too,' I nodded. 'Small world, ain't it.'

•

Bob's hangar was on the opposite side of the airport from Walt's and Joel's. Compton Airport has two parallel runways, 26 Right and Left; each three thousand six hundred and seventy feet long and sixty feet wide. Both with left hand patterns. Therefore, it depended on which side of the field you wanted to get to, as to which runway you landed on. Either runway requires a downwind leg south of the field, with either a narrow base leg for 26 Left or a wider left base for 26 Right. To land on the north side of the field where Walt and Joel were located, you had to use 26 Right. Bob's side required landing on 26 Left.

Bob and I got to be good friends and we bumped into each other now and then at various fly-ins of the Short Wing Piper Club. I remember a fly-in at Big Bear Airport one June.

•

Big Bear Fly-in

The Big Bear Airport is on top of Bear Mountain in an alpine valley at an elevation of six thousand seven hundred and twenty eight feet. A single runway aligned itself with the valley floor. The terminal building was a large cement building painted to look like it was constructed of redwood — the kind of rustic alpine airport one would expect in the mountains. It immediately reminded me of the Ahwanee Hotel in Yosemite Valley.

The Terminal Building contained the usual offices, coffee shop, and FBO on its ground floor, while its second floor was home to the Red Baron restaurant; it was complete with a flying motif of suspended model airplanes, photographs of various airplanes, and murals of airplanes. The restaurant, the entire second floor, contained a large dining room with two smaller dining rooms. Each dining room had its own stone faced fireplace and bar. It was a hopping place on a Saturday night when the 'locals' arrived to eat, drink and dance to live music.

The fly-in that Saturday brought out over a dozen short wing Pipers; four TriPacers, three Pacers, two Clippers, and three Vagabonds, as well as a couple of metal 'Spam Cans,' a Skyhawk and a Piper Cherokee.

Since the airport at Big Bear was at an elevation of almost seven thousand feet, the summer heat had temperatures on the field in the mid-seventies. Even with the warm, thin air, most of us were able to get airborne by mid-field, although our climb rates were down to one hundred and fifty feet per minute. Because of the warm air, the airplane's performed as if they were flying at an altitude of eleven thousand feet.

I remember, how at lunch, everyone seemed to enjoy the hangar flying, complete with anecdotes and vignettes of 'lessons learned' about mountain flying and the necessary safety precautions about density altitude. It was one of the more enjoyable fly-ins that year. Bob and Barbara and I spent forty minutes on the flight ramp after lunch, as we were about to leave, just trying to make the moment last.

It turned out that Bob was alone that weekend. We walked

to our airplanes parked side by side, and Bob mentioned that he wanted to fly somewhere else that afternoon before returning to Compton. He told me later that, after he took off, he flew from Big Bear to Porterville and then to Truckee, at the north end of Lake Tahoe, and back to Compton–just to see the sights on a beautiful day.

After the Big Bear fly-in, Bill Kleinbauer, one of the other Short Wingers, decided to host a fly-in for the Club at Warner Hot Springs. He was a member of the private resort that was nestled in a canyon just southeast of the Palomar Observatory.

•

Warmer Hot Springs Fly-in

Bill thought the resort would be perfect for a fly-in, since it had its own paved airstrip. For Bob and me, it meant a flight of about one hour and forty minutes each way, and seemed like an interesting place to visit, since it wasn't open to the public, except by invitation. So Bob and I decided to go. It turned out that Kleinbauer was able to host the fly-in because he led the resort management to believe that all of the Short Wing Club members were potential buyers of the resort's condominiums, as well as potential new members.

Bill told us that to be able to fly-in there, all we had to do was pretend that we were interested in getting information about resort membership. Seeming interested meant sitting through a short sales pitch upon arriving; then the rest of the afternoon was ours to have lunch in the dining room of the main hotel-like building, and then tour the grounds.

It didn't sound too terrible, so about twenty 'Short Wingers' decided to make a day of it and fly to Warner Hot Springs. On the day of the fly-in, Bob and I took off, each from our separate airports and met in the sky over a common landmark: Alondra Park, a one square mile park and golf course easily recognizable from the air, three miles south of Hawthorne Airport and six miles west of Compton.

After Bob and I found each other in the sky, which was no easy matter, we joined up and flew as a flight of two, down the coast to Oceanside where I radioed to Bob to turn to a heading of 073 degree magnetic, a direct heading to the 'Springs.' Bob replied that he'd have to follow me, as his compass needed calibrating and he couldn't rely on it. So I lead the way.

Once in the vicinity of the Warner Hot Springs airstrip, I told Bob to take the lead and I'd follow him down. As we were descending out of five thousand feet, I lost sight of Bob's airplane in the ground clutter of patchy trees and grassy areas of the parched foothills and meadows. Since I lost visual contact with him I radioed for him to forget about me and I'd meet up with him after landing.

I continued descending, and three minutes later, as I was approaching pattern altitude for the strip, I thought I saw Bob's airplane going for a 'straight in' landing, rather than following the pattern. So, without thinking or even checking the wind sock or the pattern, I began a 'straight in' approach.

I set up what I thought was a perfect landing approach, and you can imagine my surprise, as my landing turned out to be a downwind landing with a ten knot tail wind, gusting to fifteen.

Having never executed a downwind landing before, I can tell you that to call it thrilling would be a gross understatement. As I rotated to level attitude to let the airplane begin to sink, as the speed paid off, prior to pulling the nose up into a flare for a 'three point attitude,' my heart began racing, as the control wheel did little good. The airplane seemed to have a mind of it's own, and continued to alternately take off and fly from an altitude of one foot, before bouncing onto the runway and then back into the air. I couldn't get the airplane to stay on the ground. My feet were dancing on the rudder pedals and toe brakes like Ray Bolger doing a 'soft shoe' to the tune of 'Once In Love With Amy.'

Nothing I did with the controls was working. It was as if neither the rudder pedals nor the control wheel were connected to any control surface. It didn't seem to matter which way I moved the control wheel, the airplane just didn't respond, The most curious thing of all was the airspeed; the airplane was not slowing down, even with the throttle at 'Idle-Cutoff.' And since it wasn't slowing down I was quickly running out of runway.

I rode the brakes all the way down the runway as the airplane bounced into the air and then came down on the asphalt with its brakes locked I heard large chunks of rubber being gouged out of the tires. But I finally brought the Pacer to a halt, clinging to the control wheel that was now buried in my chest.

As the airplane came to a halt with its engine idling, I heard the roar of another airplane's engine at full takeoff power in front of me. I looked up through the top of the windshield just in time to see a Piper Pawnee crop duster about to crash into me head on. Seeing me at the last instant and pulling up, the crop duster was able to clear my airplane by mere inches. I was helpless to do anything but witness my imminent death.

As the sounds of the Pawnee's engine still echoed in my ears over the Unicom frequency, its pilot yelled for someone to get the dumb bastard who just landed downwind, off the active, so he could tow some gliders.

I taxied off the active and into a dirt tie down area at the edge of the runway, my heart still pounding. Everything happened so fast, I couldn't react, except for my heart racing and my breath coming in gasps. I sat in the Pacer for a long time before getting out. The adrenaline didn't stop pumping until after I'd gotten out and secured the airplane, and was walking over to where Bob and his wife Sue and their friend Eric were waiting for me.

I tied down the Pacer and thought about why airplanes land against the wind instead of with it, as I'd just done. Of course, the fact that the wind slows an airplane's forward speed over the ground is obvious. And, when an airplane lands in the same direction as the wind, the wind is not blowing over the surface of the wing as the airplane slows, since the wind's air mass is traveling in the same direction. And without the relative wind creating lift and slowing the airplane as it lands, the wind is now adding to the airplane's ground speed, by pushing it from behind. Without wind blowing over an airplane's control surfaces, they become ineffective, as stall speed nears.

I was greeted by the sounds of laughter when I approached Bob's group. They were laughing and shaking their heads in disbelief.

'What in the hell do ya call that kind of landing?' Bob yelled.

'What you've just witnessed,' I explained in an exaggerated and professorial tone, 'is a 'textbook' example of your perfectly executed downwind landing. Of course, it wouldn't have been quite as dramatic if the tow plane pilot hadn't tried to land on top of me.

'Christ, Pohlmann, are you lucky!' Bob exclaimed, he said shaking his head.

'I don't know why, but that's a fact!' I replied, wondering how many more times I would face death by doing something totally stupid.

•

I'd forgotten all about the sales pitch we were required to sit through, until I saw the resort's van waiting to take us from the airstrip to the sales office. But, after what I'd just experienced I was happy to be alive and able to listen to anything. And after the twenty minute sales pitch, we followed a flagstone path that led to the main building and the dining room where the rest of the assembled Short Wingers were waiting for us to join them for lunch.

It was an interesting place, very impressive grounds and for anyone wanting a weekend hideaway, isolated from civilization. Actually, unless you were on your honeymoon, there really wasn't anything at all to do there but bathe in the natural hot springs relaxing and enjoying the solitude. The condominiums were small single unit stone cottages with thatched roofs patterned after those of Ireland or England. They were charming, each with a cozy living room and a stone hearth and fireplace, a small dining room and one or two bedrooms, with a large bathroom. The price of resort membership wasn't cheap; but, if beauty and solitude were your thing, it didn't get any better than Warner Hot Springs.

The flight back was uneventful, and by the time I got home I'd tried to forget what had happened. But, to this day, when I see a Piper Pawnee, I am always reminded of Warner Hot Springs.

•

Throughout the years that Bob and I owned Short Wing Pipers, we shared many moments of hangar flying. Those interchanges of information and experiences were really important when one of us had a maintenance problem, especially because parts were difficult to find for certain systems on vintage airplanes. Each of us kept a card file of places that stocked certain parts.

For instance, oil coolers have to be sent to Dallas, Texas, for overhaul. In Los Angeles, there are only two places to get airplane carburetors overhauled; and there are only three local cylinder shops that can rebuild airplane engine cylinders at affordable prices. And when it comes to avionics, there is a shop at Whiteman Airpark that is good and another at Santa Monica Airport that was good, but more expensive. Through flying clubs and hangar flying, airplane owners keep abreast of where the parts are for vintage airplanes; and vintage airplanes always need fixing.

After five years, Bob moved up to a Cessna 182 Skylane, while I moved up to a Beechcraft Musketeer — both of us moving out of the 'tube and fabric' era of vintage aviation. Now Bob flies from Van Nuys, where his airplane is based. Sometimes he flies his Cessna 182 to Oregon where his relatives live. In his 182 he makes the trip in about four hours, now traveling at one hundred and fifty miles per hour rather than at ninety.

Although our first airplane was a tail dragger constructed of tubes and fabric, I think Bob Root and I experienced flying the way it was in the thirties, when it was romantic and adventurous. Contemporary pilots flying 'nose dragger spam cans' have never heard of Clippers or Pacers or Ercoupes, and wouldn't know the first thing about patching a fabric airplane. I'm kind of glad I didn't miss the fun of flying old airplanes or meeting a lot of the people that still do.

•

Bob Hayos

Bob Hayos is one of the greatest hangar flyers I've ever met. He even has a hangar at Hawthorne Airport to do his hangar flying in, and he owns an airplane. Actually, over his lifetime, he's owned many airplanes: an Ercoupe, a Cessna 170 tail dragger, a Lake flying boat, and currently a 1964 Piper Commanche. What makes Bob a truly great 'hangar flyer' is that, although he's always owned an airplane, he hardly ever flies. Instead, he mostly, sits in his hangar, on a stool next to his airplane, listening to his hand held aviation radio, and talks about flying.

He's very knowledgeable about aviation. He flew DC-3s in World War II and he's been active in general aviation and flying ever since. Through the years, Bob picked up a lot of 'tickets.' He's a licensed A&P mechanic, and he once owned his own avionics repair shop. He also worked as an electronics engineer for the Northrop Corporation for thirty years, and it's obvious that when he talks about flying, pilots listen.

The few times I've actually seen him roll his airplane out of its hangar, he'd be back in the hanger within thirty minutes. He never really went anywhere. The flights were comprised of flying a twenty mile loop, from Hawthorne Airport to King Harbor at Redondo Beach on a heading of 210 degrees magnetic, then eastward to the Harbor Freeway, then north along the freeway until abeam of Hawthorne Airport, then straight back to a landing. It was about twenty miles and took about fifteen minutes. You could usually set your watch by the time it took to complete the circuit.

I've wondered why he even bothers owning an airplane, since he only flies it to keep the engine exercised. The hangar rent alone is two hundred and fifty a month. I guess owning an airplane is an ego thing with Bob, and an excuse to have someplace to go to for serious hangar flying. And since he does his own maintenance and annual inspections, signing off the logs himself, his only fixed costs are the hangar rent and insurance.

In all fairness to Bob, I must admit that owning an airplane in a hangar is really a pain. Maybe that's why airplanes that are hangared aren't flown as often as ones that are tied down on the ramp. It's just too much work to open up the hangar doors and roll out the airplane without dinging the wing tips on each side of the hangar. And then after the flight, you have to horse the airplane into the hangar without dinging the wing tips again–not a trivial task by yourself. To make the job of alignment easier for the one-man operation owners paint lines on the asphalt in front of their hangars to line up each airplane wheel as it rolls in or out. That way, the pilot doesn't have to worry about wing tip clearance.

But as pilots get older, I think that it gets to be too much effort to horse an airplane in and out of hangar, especially for a short flight. I also considered the fact that Bob was almost seventy years old, standing five feet ten and weighing about one hundred and fifty pounds. Not exactly a 'muscle man,' so horsing his airplane in and out can be a real struggle. Since he was Hungarian, I dismissed the physical limitation concept, as all Hungarians are tenacious and if they want to do something badly enough, they'll get it done.

My analysis of why Bob didn't fly much anymore was either one or all of the following: his physical capabilities, the complexity of his latest airplane, or the fear of flying. His machismo wouldn't allow him to grow old gracefully; he had to have a similar high performance airplane to match those owned by his close flying pals.

Bob's two closest flying buddies were Mooney owners,

with airplanes that could true out at over one hundred and eighty miles per hour. And since Bob had just sold his Lake amphibian, which cruised only at one twenty, and was 'planeless' for the first time in many years, his ego forced him to buy something as fast as the Mooney. So after a two-month search, Bob bought a 1964 Piper Commanche, an airplane with constant speed propeller and retractable gear and capable of keeping up with his friends.

There was just one problem Bob had with his newly acquired high performance airplane, and that was it's variable pitch propeller and retractable wheels: the two things that made it high performance. Now to most pilots, those two features are nothing more than two more levers to move up or down or in and out, during takeoffs and landings; but to Bob, they were an extra layer of complexity that an aging pilot didn't want to think about or be bothered with. Most pilots as they get older, usually want an airplane that is simpler, not more complex.

An older pilot wants an airplane that's easy to fly and doesn't cost a lot to maintain; not one that's complicated with a heavier workload.

Most pilots in their sixties and seventies don't care about how fast or slow an airplane flies, since they've got all the time in the world to get somewhere. In an emergency, if they really need to fly someplace fast, then they can always buy a ticket and fly United.

But, I suspect that Bob's ego didn't allow him to go back to an airplane that would be comfortable for him to fly; like an Ercoupe that practically flies itself; not while his younger buddies were flying Mooneys. Besides, most pilots buy airplanes that fit their fantasies, not the realities of their situation. When a man buys an airplane he's not honest with himself about what it's going to be used for. Many buy an airplane to take those trips with the family that will never happen—the trips will only be flown in the pilot's fantasies. So keeping up with his flying buddies was what Bob's acquiring the Commanche was all about.

If a pilot is honest with himself, at age seventy, staying ahead of a high performance airplane is difficult, as the airplane is moving through the sky at almost two hundred and sixty feet per second. Any lapse of thought for more than a few seconds during a flight can be dangerous, especially through a high-density environment, like a major metropolitan area. The Los Angeles basin, a geographic area of some fifty miles in all directions, contains fourteen airports, four Airport Radar Surveillance Areas or ARSAs, and one Terminal Control Area or TCA. Most of these areas are to be given a wide berth by VFR pilots. During the last forty years, the flying environment has gotten very complex.

Therefore, most intelligent pilots who intend to continue flying till they die or can't pass the FAA physical, whichever comes first, usually give up their hot rods for a more docile and forgiving airplane. One that doesn't require a high level of skill or go fast enough to get them into trouble, if a pilot's mind suddenly begins to wander.

That's why at age fifty eight I decided not to move up to what I really wanted in a high performance cross country airplane, since I realized that in five or ten years, if I was still flying, I'd want to own just what I've got now. So why put myself through all of the aggravation of fixing up another one, the way I want it, and then sell it to someone else. All I had to do was hang on to the one I've already got.

I knew Bob had flown DC-3's during the war, so complex airplanes weren't anything he couldn't handle. But to me, the curious thing about his Army Air Corp flying was that he never talked about it. In all of his flying stories he never once mentioned his wartime flying experiences; for a pilot not to talk about himself is very peculiar. So from his silence about his military service, I got the impression that the memories were either not pleasant, or military flying was just something he was forced to do.

I don't think Bob really cared to fly fast like his friends; he just didn't want to be left behind and considered old. Speed in an airplane comes at a great cost. If a pilot is honest about it, the difference between a slow airplane and a fast one is measured in dozens of minutes, when the trip leg is three hundred miles long. To knock off a half hour on a three hour flight takes an airplane that's twice as expensive to buy and three times as expensive to maintain and insure.

•

I met Bob when I joined the Coast Guard Auxiliary in the mid-eighties. At the time, his wife, Norma, also a pilot, was the Flotilla Commander. When I first met him he was flying a Cessna 170 that he sold a year later. It seems that he'd always dreamed of owning a flying boat like a Lake amphibian, so he could fly to the various waterways of California and go camping or fishing or picnicking.

'Just think of it,' he used to say. 'Just fly into a lake and land, and then water taxi to shore and camp or tie up at a boat dock or marina. That's the life.'

Bob's problem was that his fantasy didn't belong in a California setting; it really belonged to Florida, Wisconsin, or Canada where there was a lot of water. His dream didn't square with the geographic realities of the Golden West. Most of the water in California, besides the ocean, is found in reservoirs; and most of the reservoirs cater to fishermen and boaters; not seaplanes. Local governments really try to attract the boater or fishermen because they don't just fly in and out; they stay over the weekend spending their money on the local economy, like motels and restaurants.

Reality certainly interfered with Bob's dream–but what about the ocean? There's plenty of water there. Actually, the ocean doesn't really count for a place to land a small float plane or flying boat, became it is normally too rough. The Lake amphibian was designed to land in fresh water lakes, where waves never get over a foot and a half high. But the final coup to Bob's fantasy was that California, at the time Bob considered buying a seaplane, was in the third year of an extended drout, with no end in sight.

During wet years, when water levels were highest, there were only a couple of lakes of any size that allowed seaplane operations, and those were all located north of San Francisco. Places like Lake Shasta, Lake Tahoe, and Clear Lake. But none of them were real lakes. They were all reservoirs that were dammed up and whose water level was always a function of how much water the farmers of the area were buying from the County or State.

Seaplane bases were symbolized on the sectional charts for the big three lakes; while the remaining reservoirs that still contained enough water for boating, were restricted from seaplane usage.

Actually, nobody knew just exactly how boats and seaplanes would mix in the same water. I was suddenly reminded of Bob Johnson's problem with his float plane at Bass Lake in Ohio from my youth. I remembered the trouble Johnson had with the fishermen of the lake in 1948, when he tried to share the water with boats. Bob Hayos had the same situation — trying to mix fishermen and seaplanes several generations later.

But, a dream is a dream, and reason and logic have no place, when it comes to fantasy. In the end, it was Bob's Hungarian stubbornness that forced him to buy the airplane of his dreams: the Lake amphibian. His dream would not be denied. After all, he'd already told everyone about his fantasy, ensuring that it would become 'self fulfilling prophesy.'

Bob didn't look Hungarian, although he was, and he was fluent in the language. Nothing about him looked 'Slovakian.' He looked like an Englishmen, with a face so white he looked like he'd never been out in the sun. He would've made a great spy; as no one would ever suspect him of still having relatives in the 'old country.'

His complexion was fair, his stature average and his salt and pepper hair was worn short in a crew cut — probably a throwback to his military days. Everything about him said 'English,' except his temperament; that was certainly Hungarian. His temperament never really surfaced until he decided something, and then nothing short of death could turn him from his chosen course of action.

He told me some time later that he bought the Lake because he always wanted it, and, because of the Coast Guard patrols; but mostly because he'd always wanted it. And as I thought about it later, I didn't think his rationale concerning the patrols made sense; as it was my contention that a small, under powered four place flying boat really wasn't safe when flying over the open ocean. The air patrols of the Coast Guard Auxiliary were usually flown from the coastline up to ten miles off shore, except for the North Sector, where the patrols were flown up to forty miles at sea, off shore around the Channel Islands, west of Ventura.

If a Lake flying boat lost its engine at a patrol altitude of five hundred feet above the water, and the waves were less than two feet in height, with no wind, the landing would be 'iffy' at best. Landing a flying boat in the open ocean would be easier than landing a float plane, but with waves normally two to three feet high, even the flying boat might flip over on landing and sink. Most general aviation airplanes that I've ever heard about that ditched, whether in the ocean or a lake, only remained afloat between seconds and one minute, before sinking. And, it didn't seem to matter whether they remained right side up after landing.

•

Every year the United States Sea Plane Pilots Association holds a west coast 'Splash In' at Clear Lake, California. Although he didn't have his seaplane rating yet, Bob had joined the Association years before, and had followed their fly-in events for a long time. For as long as he could remember, he'd dreamed of attending the 'Clear Lake' thing in his own seaplane. He wanted to know the joy of a water flyin, and a weekend in northern California on a large lake in the company of seaplane pilots was the next thing to heaven.

Actually, for all geographic areas south of Vancouver, the seaplane 'Splash In' at Clear Lake is pretty much the only seaplane thing going on in the west. And it was now the third year of a six year drout, and all of the reservoirs in California were practically dry. The water level at the San Luis Reservoir, near Los Banos, was down over fifty feet. What was left of the water was now just a single channel two hundred feet wide and a half mile long. The channel was right in the middle of the reservoir which now looked like a large dirt bottomed basin. Normally, the reservoir, took more than a half hour at high speed to drive by in a car. It was a really big hummer — almost as large as Clear lake.

Even Lake Isabella, nestled in the center of the Sierra

mountains, east of Bakersfield, showed more lake bed than water. In fact, the level was so low that the old town of Isabella, normally submerged when the lake was full, was now beginning to surface, like a submarine. The skeletons of trees and the remains of the town's water tower could have seen above what water remained it's level down almost eighty feet. Water was really in scarce supply in California, with statewide rationing in effect even in restaurants. It was so serious, that neighbors would report to the police anyone guilty of washing a car in the owner's own driveway.

Bob had fantasized about the flight to Clear Lake for months after he'd sold his Cessna 170 and bought his beloved Lake flying boat. He found a CFI at Santa Monica Airport that offered lessons for the seaplane rating and could check him out.

The week before his flight to Clear Lake, Bob could think of nothing else; he'd planned and replanned the trip both ways. He had a feeling that it was going to be a great adventure. He visualized himself with Norma, next to him, as they flew northward. He had even outfitted the flying boat with a stereo system so that, as they cruised along, they could listen to their favorite music over the intercom. The visualization was wonderful and not only kept him motivated and anxious for the day of departure, but also kept him from considering the realities of his airplane's actual performance.

He was so in love with the Lake flying boat and with the idea of the trip that he never thought about the airplane's performance limitations, as he planned the flight. According to the Pilot's Operating Handbook, the airplane should cruise at a speed of one hundred and thirty miles per hour. And that seemed reasonable since it had a two hundred horse engine. But, the realities of flying boats are that even the fastest of the Lake series of airplanes, the Renegade model, with its two hundred and fifty-horse fuel injected engine, still cruised at only one hundred and thirty five miles per hour. Only the turbocharged Renegade was able to fly one hundred and fifty miles per hour; but unlike Bob's forty thousand dollar airplane, the Turbo Renegade cost a quarter of a million. What price speed? For that extra twenty miles per hour, the airplane's selling price would cost an additional two hundred and ten thousand dollars.

Needless to say, Bob never considered any of this; he just assumed that if the book said one hundred and thirty, that was the speed it was supposed to go and that was the speed he used for the flight planning. Bob just never gave his fantasy the acid test–reason, and the laws of physics.

The seaplane base at Clear Lake is a distance of four hundred and seventy two miles from Bob's airport, a little too far to fly non-stop for the Lake's range. Bob would have to refuel after about four hours, especially when flying northwest, the usual head winds would slow his ground speed some. But that was okay, he thought, since they could land at Watsonville on the flight north and at Santa Ynez on the flight back. After all, at a cruise speed of one thirty, it would should only take a little over three and three quarter hours of flying time, and maybe an extra hour to refuel.

The Hungarian in him dismissed the thought that sitting in a small airplane for hours without moving, except to squirm, was to incur rigor mortis by the end of the flight. After all, this was his dream and he never considered for a moment that it might not be Norma's dream. He did wonder, however, why she never once came to the hangar to look at his newly acquired flying boat, the way she had each of its predecessors. And that she never once talked about getting checked out in it as a pilot, the way she had the others. He didn't understand why she was ignoring his airplane, but he knew that once they were airborne on their trip, Norma would fall in love with the Lake, the way he had.

Now, for anyone who has flown in small, general aviation sized airplanes, a flight of more than several hours can pretty much stiffen one's body. Of course, these realities of flight were never a part of Bob's fantasy. And Bob also never considered that if they didn't get an early start on their return trip home, the fuel stop might extend their flying into the evening hours and maybe even into darkness. Night flying was definitely not in his fantasy. His was a daytime VFR flying fantasy, only.

To complete reality's acid test of Bob's flight planning, it should be noted that a flying boat with an engine suspended on a four foot pedestal, above the fuselage, equates to a lot of drag. It could probably slow the airplane from the book figures by ten to fifteen miles per, making the actual ground speed about one fifteen to one twenty.

I would also note that when flying towards San Francisco, north of the Santa Barbara area, there's a twenty mile per hour head wind for most of the trip. Rendering the two hundred horse powered Lake, a real world ground speed to Clear Lake from Hawthorne, of a measly one hundred miles per hour. This equates to a flight of almost five hours, with another hour for the enroute fuel stop.

When the day came for the trip, undaunted, Bob and Norma launched on Bob's 'perfect adventure.' And, at the end of the day, when the flight was over, it had taken nearly six hours of flying time. So that, by the time Bob and Norma climbed out of the Lake and stood up, their bodies felt like pretzels.

The airplane seemed to take forever to get there. Its cruise speed calculated out to about one hundred and four miles per hour, a whole lot slower than the 'book' promised. By the time Norma got out to walk around at Watsonville Airport, their fuel stop, she wished that she was already at the destination. Her entire body ached and was so stiff, she could hardly move.

Bob should have realized that the trip wasn't all that wonderful, since at the fuel stop, Norma, who usually was very talkative, stood stoically silent, as line service pumped fuel into their airplane. When Bob asked her if there was anything wrong, she was terse with her reply about how long it was taking to get there. She didn't think it should take all day.

The flight from Watsonville to Clear Lake took an hour and a half. As they descended to the lake, Norma's spirits buoyed, as she realized that in minutes she could climb out of their airplane and walk around. When Bob saw all of the seaplanes moored along the lake's edge in front of the motel, his spirits climbed as well. And once they were down and the airplane was moored and they were surrounded by dozens of other seaplane pilots wishing them well, Bob was at peace with the world. Norma was just glad to get out of the airplane.

•

The weekend went by too quickly, but there were, as advertised, flying games of water bomb contests and spot landings, as well as the tutorials about water flying. Bob hadn't entered any of the contests as he hadn't been able to fly his airplane enough to learn its performance, having just purchased it prior to the fly-in.

At the Saturday night steak fry, Bob and Norma made many new friends and looked forward to other 'splash ins.' Everything was going great, just like Bob's fantasy, until the next morning, when it was time to go.

It was going on eleven o'clock after the pancake breakfast and time to go. Bob loaded their luggage into the flying boat, as Norma said her good-byes to some new friends.

When Bob finished loading the airplane, he water taxied it to the fuel dock, shutting down the engine while waiting for his turn to refuel. After an extended wait, it was finally his turn, and he patiently sat in the cockpit of his airplane with the cabin door open, as the line boy filled his tanks.

As he sat there, a Cessna 172 on floats taxied towards the fuel dock and shutdown, coasting directly towards Bob's seaplane. Bob wasn't paying attention much, as he sat there enjoying the moment. He looked up just in time to see the Cessna heading straight for him.

The other pilot tried frantically to steer the float plane away from Bob's flying boat, but the speed of the floats had fallen off, making the water rudders ineffective. The Cessna plowed its starboard float into the right pylon of Bob's wing float at three miles per hour, denting Bob's pylon float.

In the end, the accident cost Bob almost five hundred dollars. And the incident ruined the memories of Clear Lake. It would be one 'fly-in' he'd never forget.

•

One Sunday afternoon, after the 'splash in,' Bob and I were sitting in his hangar and he looked over at me and said, 'Ya know, it's a funny thing. I dreamed of owning a flying boat for twenty years. And now, after I own one and even took it to a fly-in, I've kind of got it out of my system.'

'What do ya mean? Ya want to sell it?'

'I know it sounds crazy. But dammit! It's just too slow. It took us all day just to fly to Clear Lake from here. Ya'd think a two hundred horse engine would move faster.'

'But look at the engine, Bob,' I said, pointing. 'It's sitting on top of a four foot pylon over the fuselage in the airstream. That's a lot of drag to overcome. Besides, there are other places to fly to that must be a lot closer than Clear Lake.'

'Not open to seaplanes, there aren't. Most of the reservoirs are still restricted to just boaters.'

'Hell Bob, with this drout, there's no water in any of 'em anyway. Even the boaters are giving up,' I said, looking over at his airplane, behind me. 'It sure's a pretty airplane.'

'I dreamed about flying a Lake for twenty years,' he rambled. 'It just never went fast enough for a long trip. Coming back it got dark on us over Santa Barbara. The last hour was flying at night. Norma hates flying at night.'

Two months later, after owning the Lake for only eight months and only flying it to one 'splash in,' Bob sold it to a retired Airline pilot, who had seen it at the Clear Lake fly-in and thought it would be fun to own and fly to various lakes. Bob didn't say anything. He just nodded, understanding perfectly another pilot's dream.

The retired airline pilot paid cash.

•

Earl Boone

I first ran into Earl Boone in meetings at work, where he was a specialist in military marketing. It was months and many staff meetings together before I got to know much about Earle. In some ways, he is very open, but in others he's very private. A person only gets to know as much about Earle as he wants them to know — a habit he probably picked up doing 'odd jobs' for various clandestine agencies of the government, during his years in the military.

Earle's a sixth generation descendant of Daniel Boone. And I can see how his namesake, way back when, could do all of the things he did. I saw in Earle the attributes that it takes to motivate a person to 'be all that they can be.' Earle is very much like my 'mountain man' friend, John Kieth: self reliant, a self starter, and a loner. Like John, Earle would never pay someone else to do anything that he could himself.

After knowing Earle for a long time, I finally decided that he is a 'Type A' personality, and although he's soft spoken, with a gentle southern charm that bespeaks his heritage, his approach is still quietly aggressive and persistent.

Earle not only knows about everything, but he's current on everything. And by that, I mean he stays current on news, politics, business, arts; you name it, and he can converse intelligently about a subject at length. He is a veracious reader and observer of world events, watching several news and current events programs on TV that are acclaimed for their objectivity and accuracy in reporting. I don't think that he ever watches TV for entertainment. I wonder is he ever does anything to relax?

What I find most interesting about Earle, is that if he doesn't want you to know something about him, you'd never find out: that he's a retired Air Force Colonel, that he was a pilot and flew fighters and bombers, that he ended his career in the military by being chosen to be part of the select group of pilots that flew the SR-71, or that after he retired from the military, he was a corporate pilot for a King of one of the Middle Eastern countries.

•

When it came to hangar flying, Earle could talk about things that not only challenged a pilot's skill level, but challenged a pilot's character, as well. To Earle, everything in life was a challenge. If a task was too easy, Earle would find a way to do it in a way that was more challenging. One day at work, he and I were hangar flying about B-58 Hustlers during Viet Nam. Now normally I'm not interested in 'warbirds,' but since I'd worked at Douglas Aircraft Company where the airplanes were manufactured, I happened to know something about them.

I got interested in talking about the airplane with Earle, because he knew the other half of the story, from the Air Force side; since the airplane was used by both the Navy and Air Force.

The Air Force B-58 was the same airplane that the Navy designated the A-3D, usually referred to by Navy pilots as 'all three dead!' The somber acronym attested to the fact that the Navy version of the twin jet bomber wasn't outfitted with ejection seats, using instead a tunnel-type chute in the bottom of the fuselage where the crew was supposed to 'slide down and out,' to escape from a crippled or burning airplane.

Anyway, during his military career, Earle had flown the B-58 and now, while hangar flying, he was telling me about the fact that the airplane was so unstable in flight, that a computer had to augment the flight controls. Never to be intimidated by a machine, Earle took the airplane's instability on as a personal challenge. To Earle, flying the airplane manually, without control augmentation had to be mastered, if he was going to survive in combat as its pilot. After all, he reasoned, in combat the flight control augmentation could be damaged, and if he never trained for that contingency, then that system failure might be the 'single point of failure' that would cost him his life.

Earle was probably the only pilot in his squadron or wing that ever practiced flying a B-58 without the augmenting system engaged.

'If it's so difficult to fly manually, how could ya practice?' I asked, wondering.

'Well towards the end, I used to fly maybe half a mission with it on manual,' he said, grinning. 'It took a long time for me to work up to that point, but I took it on as a personal challenge.'

He told me one time that because he tried to be the best pilot in his squadron, he was given special assignments. Successful in these assignements he was selected to fly the SR-71 'spy plane.'

While Earle's contemporaries were pushing an airplane's 'envelop,' Earle was pushing both that envelop and his personal 'envelop' as well.

•

From the time he was a kid, Earle's love affair was speed! Speed is Earle's thing, an intoxicant. As a kid, I could visualize him trying to go sixty miles per hour on his bike. Of course, flying jet airplanes was a natural progression for him, as well as a 'natural high.' He just loved going fast.

Before he retired from Northrop and gave up commuting fifty miles each way, I asked him one day how he liked his long drive. He said he didn't really mind, because he made a game out of it.

'What do ya mean?'

'I see how fast I can go.'

'On Los Angeles freeways during morning rush hour?' I laughed, thinking about all of the traffic. 'So how fast can ya go in the morning?'

'Eighty!'

'Eighty! How the hell can ya go eighty with the traffic bumper to bumper?'

'Well, first you have to get on the freeway early in the morning,' he began, sounding very clinical. 'Before the traffic gets too heavy; usually before sunrise. Then, you have to give the act of driving your complete attention. And when I say complete attention, I mean one hundred and twenty per cent concentration. So much attention that nothing that's behind you or beside you can distract your concentration from the conditions ahead.'

'Well, what happens if the car to your right swerves in front of ya suddenly?'

'I move onto the shoulder area between the fast lane and the center divider. That's why I only drive in the number one lane, so I always have an out.'

So Earle would drive the fifty miles to work each morning, actually enjoying the challenge.

•

A year later, after Earle had taken an early retirement, Barbara and I spent the fourth of July holiday weekend with Earle and his wife in their new home in Scottsdale, Arizona. By noon, the temperature was one hundred and ten degrees. Earle wanted to take me for a ride in his pride and joy 'kit car,' that he'd built some years before. It was a replica of a 1929 Mercedes Benz SSK, and it looked very much like a 1952 MGTD I'd owned in the late fifties.

It was small and I had to carefully tuck myself into the passenger seat. Earle had taken the convertible top off, so I could get the full effect and sensation both of speed and I supposed the one hundred and ten degree temperature.

He slowly and carefully backed the car out of his garage and started up the meandering streets of northern Scottsdale. We sped along at fifty until we reached Pima Road, a large country boulevard that passed through the undeveloped area of the city. It cut through the open desert, running north and south all the way down to Phoenix, some twenty miles away. Normally, there never is much traffic on it, and in today's heat, there was no traffic at all.

Earle pushed the accelerator to the floor.

I was expecting about the same acceleration that was in my under powered MG, which was not very much at all. But I wasn't ready for what Earle had under the hood.

As soon as he 'floored it' our heads snapped backwards with a sudden jolt! Our bodies were pinned against the leather seats and I looked over at the speedometer; the needle was moving past ninety and closing on one hundred.

'Wow, Earle!' I gasped. 'What/s under the hood?'

'Ford V-8, he grinned, quickly glancing in my direction before returning his fixed gaze on the road ahead.

'How fast does it go?' I asked, hunkering down in the seat, as the wind pushed against my face uncomfortably.

'It's designed to go one hundred and ten miles per hour, and that's exactly as fast as we will be going in another couple of seconds,' he said, as his eyes darted to the speedometer and then back to the road.

I looked over and sure enough, the needle was sitting on '110.' I looked ahead at the road and wondered what would happen if we ever hit something or blew a tire going this fast. They'd never get our bodies out of the 'balled aluminum.'

The car almost floated as it would come up to the top of a gentle rise, and then sink back down onto the road, as the road leveled out. I began considering the possibilities of actually turning over and careeing off the road into the open desert.

There was no roll bar. And this wasn't like driving my MG, since it was never capable of going fast enough to blow a tire. It usually took a typhoon just to get it moving. And the only time it might go eighty would be straight down over a cliff!

We drove along for several miles at top speed as Earle just sat there, grinning, enjoying the sensation of speed once again. Finally, we began to slow down to sixty. And at sixty it felt like we were standing still. And by the time the ride was over, two things had happened; Earle had his speed 'fix' for the day, and I'd been pretty much been scared to death.

•

Bob Wenholtz

It is summer, and the weather is very hot for Los Angeles; the daytime temperatures are in the low one hundreds. It's Saturday, but too hot to fly. I went down to the airport to check on my airplane, and just finished checking out its cockpit and was securing the cabin door, when I heard noises from the end of the parking ramp, four airplanes away. I looked up and saw a small black girl wearing a black jumpsuit, washing the Cessna 210 at the end of my row.

She wore her hair up in a ponytail because of the heat. I looked at my watch, It was noon and very hot, and I thought to myself as I watched her for a moment, that she must be really uncomfortable in that black one piece jumpsuit–exactly the wrong apparel for this weather. After a moment she climbed under the airplane, scrub brush in hand. As she worked, I thought about the reason I began renting a parking spot from this particular FBO; it was mostly because of their low monthly rates, but also because they advertised a free airplane wash every month. And at the time, an airplane wash on the field cost about fifty dollars.

The free wash was what really convinced me to move my airplane from the City owned tie downs to those of this FBO. But, in a year of parking there, my airplane never got washed, not even once. The FBO couldn't seem to keep the people they hired to wash airplanes. Now, suddenly seeing a new 'washer' actually washing an airplane in my row made me curious enough to stick around to see if she'd end up washing mine.

After I locked my airplane, I got into my car and drove behind another row of airplanes, where I could wait and watch, parking in the shade of a Cessna's wing.

Two airplanes away from my car was an elderly man puttering inside his vintage Cessna 182. Having nothing better to do while waiting for the girl to finish and move on to the next airplane to be washed, I began studying the finish on the old man's airplane.. It was obvious that it was in need of paint, but it was also obvious that someone had spent real money on it in the way of modifications. It had stall fences installed on top of the wings reminiscent of the Robertson STOL kit, to improve its low speed controllability during landing.

From other flyers that parked at the FBO, I'd learned that the 182 I was studying belonged to a retired Continental Airlines Captain, who busied himself in retirement by flying doctors and dentists down to the lower stretches of the Baja peninsula, where they would treat poor villagers that had never been to a medical clinic. The doctors were all volunteers and members of the Flying Samaritans. They were flown down to Mexico for three days at a time to treat the poor. It was kind of a mini-vacation. They would treat the sick for a couple of days and then would do some serious fishing before being picked up for a return flight. And as I remembered, the 182 was mostly gone during week days and only there for one or two days on the weekends, before disappearing again.

That kind of bush flying was tough on an airplane, and I could see why he'd never bothered to paint it. I also noticed that the wheel skirts were removed from all three landing gear. That's because on a dirt strip that may have ruts, skirts have a way of being ripped right off of the wheels during rollout. Also, 'urban camouflage' helps protect an airplane from theft, when the outside paint makes the airframe look 'ratty,' and not airworthy.

As I sat in my car studying the 182, its owner suddenly appeared outside. As he emerged from the cockpit, he also studied the girl airplane washer. The next thing I knew, he was crossing the rows of parked airplanes, walking over to her. As he passed my view, I saw that he was probably close to eighty years of age, although he had a very spry step. For several minutes he and the girl conversed with animated movements, motioning in the direction of his airplane.

The girl returned to scrubbing the underside of the 210, and he turned and walked back towards his airplane. But, as he passed in front of my car, he spotted me and stopped by my open driver's side window.

His crows feet were the tacit reminder of the thousands of hours he'd spent squinting at the instrument panel. His sparse hair, was a sun bleached blond. He was of average height and medium build actually in great shape for his age, which he finally admitted was eighty three. In a Captain's uniform with lots of gold braid, he would have been the Madison Avenue image of the perfect airline pilot or Navy Admiral.

I could tell right away that he was very bright and quick, his speech patterns modulating between slow and deliberate, fast and descriptive. His phrasing was precise and elegant, evidence of his apparent education and appreciation of the language. He was definitely a 'Varsity Player,' and reminiscent of Earle Boone.

For twenty minutes he went on about all of the things that were wrong with the policies and operation of the FBO. Finally, I got out of my car and stood in the blazing sun, as he continued talking. As hot as it was neither of us moved. He was intent on telling me what he knew about the operations of the airport; and he knew a lot. After a half-hour, the conversation shifted to his 182 and his work with the doctors. It seems that he flies down to Mexico so that medical aid can be brought to the 'poorest of the poor.'

You talk about a great 'hangar flyer,' man he was it! I didn't move; just stood there listening intently to what he was saying, despite the searing heat. Occasionally I'd nod and grunt, just to let him know that I was following his thoughts which ranged from the history of the FBO in front of us, to the history of the Airport.

It was like talking to Lindbergh, as an old man; he'd been everywhere and done everything, and at eighty three was still flying back and forth to Baja at least once or twice a week, logging fifteen to twenty hours per week. And from the looks of his airplane, most of his flying was done in the bush.

He started out talking about his teenage and college years–it was almost like listening to 'Volume I: The Early Years,' or 'An Evening With Bob Wenholtz.' or Garrison Keiler of 'Lake Wobegone' fame.

He talked for two hours, while I listened. But, after all, two hours in the broiling sun was just about all either of us could tolerate, and since it was evident by this time that the girl had left and had no intention of returning, we ended the conversation and said our goodbys.

As I drove out of the airport, I thought about one of the most interesting things that he'd said to me, which happened to be about his wife, Helen. It turned out that she was quite a pilot in her own right and had participated many times in the National air race for women, the Powder Puff Derby, as well as the Palms to Pines race, from Santa Monica Aiport to Independence, Oregon. Several times she'd won her class, flying the same Cessna 182 that Bob now flies south of the border.

If Helen was half as verbose as Bob, I'll bet her flying tales would be just as fascinating. They could market 'An Evening With Bob and Helen Wenholtz' as a two volume set of CDs; maybe an album.

•

The curious thing about pilots and airplanes is that although Bob's airplane is parked just sixty feet from mine, it might be a week, a month, or a year, before the two of us might run into each other again. Half the time when I arrive at my airplane to go flying, Bob's plane is already gone and in its place is his pickup truck with the camper shell on its back.

Chapter Nine
Weekday Flying

While Waiting For The Weekend To Come

...The moonless night gave no hint of the geography below. I carefully banked the single engine Otter as it intercepted the 133 degree magnetic heading of the Paso Robles VOR station and began tracking its electronic beam. The airway designated Victor 25 turns slightly southwest from King City, and then heads straight for the San Marcus Pass, eighty one miles ahead in the darkness.

I checked the directional gyro after the bank and then glanced at the chart, verifying the new heading. Satisfied, I returned my gaze to scanning the instruments.

A few feet in front of the instrument panel, the 650 horse Pratt and Whitney R1430-S1H1-G engine droned on monotonously, its powerful thrust pulling the heavy airplane slowly through the cold night air. The engine periodically sent waves of vibration through the airframe. I could watch the visual effect of the vibration as tiny ripples were formed on the surface of the half filled cup of coffee, as it moved slowly to and fro in its gimbaled holder, above my left elbow. There was just enough glow from the post lights that illuminated the individual instruments of the instrument panel to see the ripples form on the surface of the coffee.

The cockpit floodlights bathed the instrument panel in a red glow, like a photographer's darkroom. I adjusted the propeller control and then glanced down at the performance chart in my lap. With a flashlight in my right hand, I compared the numbers indicated of the engine gauge with those of the chart. Satisfied once again, I put away the flashlight and looked out the side window.

With no moon, there was nothing to see. The night sky and the ground below were fused into total darkness. But every few minutes a single point of light from a solitary ranch below announced itself from the blackness. The single light shined like a beacon; just powerful enough to pierce the night. After a long moment I resumed my instrument scan.

The LORAN C that measures the distance between two stations was continually tracking the airplane's progress to San Marcus. Its indicator slowly changed the nautical 'miles to go.' — first, '74,' then '73,' then '72,' as the distance lessened each moment. As the numbers approached the 'sixties,' I realized that the airplane was now flying over an area of the Sierra Madre foothills where on another moonless night, a small twin engined airplane had recently crashed.

The twin was carrying a cargo of U.S. Mail from San Francisco to Los Angeles. It was a Model 18 Beechcraft from World War II. It had a problem with its electrical system, and lost all ability to illuminate its flight instruments.

Even though I had just used the flashlight, I pulled it out of the map pocket to check it again, turning it on and off several times. I remembered that the twin Beech went down on a night so dark that no horizon was perceivable, and in total darkness, its systems had failed it, instantly immersing the cockpit into complete blackness.

Of course, pilots, by law, are supposed to be prepared for

emergencies, especially when flying at night. But that night, as the pilot reached for his flashlight, he discovered that it wasn't there. He frantically searched the cockpit for it and several other backup flashlights, but none were found. Then he remembered: during the preflight in San Francisco, he'd been rushed and was behind schedule, skipping certain items on the checklist, like the cockpit area.

Without a flashlight to see the flight instruments the pilot realized that he was in serious trouble. He tried his best to keep the airplane straight and level, but it was impossible. In total darkness, with no frame of reference, he had only the sensations of his body to guide him. There was no way to tell up from down, and he knew it. One of the first things an instrument pilot learns in training is never to trust his body's feelings when flying instruments. Only the flight instruments tell the truth without emotion, as long as they are working. No matter the situation, the instruments are unaffected by fear or death; they're always right. But, to interpret them and to understand what they are communicating, the pilot must be able to see them.

The twin Beech slowly banked to the left in a descending spiral, imperceptibly at first. The airplane was equipped with propellers that maintained a constant speed regardless of the airplane's attitude. There was no high pitched scream as the engine's wound up and the airplane went into a dive. They just continued to drone on, unaffected, as if in level flight.

There were no aural cues to alert the pilot of an attitude change, and no visible cues of a horizon. Suddenly, the pilot realized from his body's weight against his seat belt, that the airplane was in a 'graveyard spiral,' that would terminate in his death if he couldn't intervene.

Frantically, the pilot radioed his plight to Los Angeles Center, where the radar controllers were already monitoring the Beech's erratic maneuvers on their scopes. They already knew better than the pilot that something was terribly wrong.

The pilot just had time for the one quick radio call before the airplane became inverted and dove straight down into a dry wash, seventeen miles west of the ranch town of California Valley. As the airplane plunged, the pilot continuously tried to level its wings.

At the high desert town of Palmdale, 150 miles to the east, at the Air Route Traffic Control Center, five controller's huddled around the Sector Controllers radar scope, after receiving the one frantic radio call. Helpless, they watched as the airplane's altitude numbers fluctuate on the airplane's data block on their screen.

First, the blip of light moved slowly off course, and the altitude numbers on the screen next to the blip increased by one hundred feet, and then decreased by five hundred, before finally counting backwards to '21' and then disappeared from the screen.

The blip that was the Beech 18 never returned to level flight. The altitude numbers of its data block suddenly told the story; counting backwards by hundreds of feet until reaching '21,'indicating two thousand one hundred, the elevation of the foothills where the airplane went in. It was less than a minute before it was out of control.

The radio calls of the controller to the Beechcraft went unanswered.

To think that forgetting something as trivial as a flashlight would cost a pilot his life. That was a scary proposition. I cycled the flashlight on and off several times, just to make sure it would work, before replacing it within easy reach.

My eyes fixated on the airspeed needle, and for several seconds I stared at the needle positioned on the number '128.'

BLEEEEP... **BLEEEEP**... **BLEEEEP**... the stall warning horn suddenly came to life, blaring like a trumpet.

Suddenly I jerked awake!

At first I didn't understand what was happening. The horn blasted again. Then the sound of screeching brakes filled the air. It wasn't the stall warning horn after all, but the car in front of me and behind me, as the traffic in the fast lane of the 105 Freeway suddenly stopped.

I'd been daydreaming about flying again. Here I was driving home from work and daydreaming. I wondered if other Sunday Pilots ever daydreamed about flying while passing the time during their freeway commute?

The End

Glossary of Aeronautical Terms

What Did He Mean By That?

The following glossary may be helpful to non-flyers in understanding the aviation terms referred to in this book. Sunday Pilots might find this glossary useful to see how many terms they remember since their last ground school or Biennial Flight Review.

A-4 — A single seat jet fighter aircraft built by Douglas Aircraft and named the A-4 Skyhawk. For more than twenty years, it was the U.S. Navy's smallest carrier-based jet fighter plane. It was often called 'the world's smallest atomic bomber.'

ADF — Automatic Direction Finder. A low and medium frequency radio receiver sometimes referred to as a radio compass, since the needle of its indicator points in the direction of the incoming signal, making it useable for navigation. Since the radio can be tuned to commercial broadcast stations, as well, Sunday Pilots use it to listen to music or baseball games (when they probably should be navigating by it).

AERONAUTICAL CHART — A map used in air navigation that portrays symbolically to scale, topographic features, hazards and obstructions to flight, as well as the locations of navigational aids, airway routings, designated airspace, and airport locations. The two commonly used charts for general aviation are the Sectional Chart and the WAC or World Aeronautical Chart. Sometimes, Sunday Pilots are guilty of using Auto Club road maps of the state highway system for inflight navigation.

AIRMET — Airman's meteorological information. An inflight weather advisory that updates an area forecast concerning weather that is dangerous to Sunday Pilots, such as icing conditions, turbulence, high winds, or low ceilings and visibility. The kind of weather that only medical doctors 'would be caught dead in, as they fly their expensive and high performance Beechcraft Bonanzas.

AIRPORT ADVISORY — An advisory of the landing conditions at an uncontrolled airport (one without a control tower), usually broadcast over a radio frequency used for that purpose alone, called Unicom. The advisory includes the current wind conditions, altimeter setting, and the runway in general use. Sunday Pilots use Unicom to find out if the restaurant on the field is open, and whether they serve beer or wine.

AIRWORTHINESS DIRECTIVE — A notice from the Federal Aviation Administration to the owner/operator of a specific airplane model (designated by serial number of engine or airframe), that an item of maintenance or condition exists that if not corrected by a prescribed time or date (sometimes before the next flight) creates a dangerous flight condition, making the airplane not airworthy. For Sunday Pilots this means, that if an AD isn't complied with, and the airplane crashes on the next flight killing the pilot, the underwriters aren't liable and the insurance doesn't have to pay for the damages.

ASR APPROACH — An approach to landing at a controlled airport that uses the Airport Surveillance Radar (ASR). ASR is normally used by Approach Control, Departure Control, and tower personnel to provide vectors (magnetic headings for the pilot to fly), as well as aircraft separation within the immediate vicinity of the airport. Sunday Pilots normally use this service when landing, as they often don't know where they are while looking for their destination airport.

AIR ROUTE TRAFFIC CONTROL CENTER — Sometimes referred to as 'Center,' is a facility providing Air Traffic Control (ATC) services to airplanes operating on Instrument Flight Rules (IFR) Flight Plans, within controlled airspace. The Center normally controls the enroute phase of an instrument flight.

ALTERNATE AIRPORT — An airport at which an airplane may land if a landing at the intended destination airport becomes inadvisable because of weather conditions. For the Sunday Pilot, an alternate airport is whatever airport he lands at, in case it isn't his intended destination when he thought it was.

APPROACH CONTROL — An air traffic control facility controlling aircraft either arriving or departing a terminal area of an airport. For Sunday Pilots, Approach Control is usually the agency they call on the radio when lost, low on fuel, and out of ideas, while trying to land somewhere.

APPROACH GATE — An imaginary point, usually one mile from the Final Approach Fix of an instrument landing system approach to an airport. It's the 'funneling in' point that Sunday Pilots are herded into before Approach turns them lose to execute an instrument approach on their own.

AIRPORT TRAFFIC AREA — An five mile radius from the center of a controlled airport. Pilots are to receive a clearence by the tower prior to penetrating this area. It is also the area that Sunday Pilots fly into without permission, and then have to apologize profusely to the tower for executing such a dumb act.

ATIS — Automatic Terminal Information Service. The continuous recorded message that is broadcast over the radio, relating routine but essential airport terminal information. By recording the data, the controller's workload is greatly reduced. A separate alphabetical designator is assigned to each broadcast, so more current messages can be identified. Such a broadcast may sound like: 'Santa Monica Airport information 'Yankee.' One three zero zero Zulu. Weather; measured ceiling two thousand overcast, visibility three, haze, smoke, temperature seven one, dew point five seven, wind two five zero at five, altimeter two niner niner six. Runway 25 approach in use. Advise on initial contact, you have 'Yankee."

ATP — Airline Transport Pilot Rating. It is the rating that all airline pilot's are required to attain to qualify for the rank of airline captain. Sunday Pilots normally get this rating designator confused with something you put in the crankcase of their car as an oil 'additive.'

BASE LEG — The Base Leg of an Airport Traffic Pattern. A flight path at right angles to the approach end of the landing runway; extending from the downwind leg to the intersection of the extended runway centerline. If that sounds confusing to you, just think how difficult it is for a Sunday Pilot to fly, while try to figure this all out.

BEECHCRAFT MUSKETEER — A four place, single engine, low wing monoplane of bonded aluminum construction. Originally designed as a trainer, it is one of the roomiest of all airplanes in its class. But, because of its greater cabin size, it is also the slowest airplane of its class. Although most of the parts are fabricated out of aluminum, steel and brass, when Musketeer mechanics order replacement parts from the factory, the price indicates that all of the parts are gold plated before being shipped. Sunday Pilots equate owning any model of a Beechcraft airplane with a bottomless hole in which to throw all of their life savings just to keep the airplane operational.

BEST GLIDE SPEED — The speed at which an airplane in a particular configuration, usually after the engine quits for good while in flight, miles from the nearest airport, is able to glide the furthest with the least amount of altitude loss. When Sunday Pilots are island hopping and lose an engine while enroute, it is the speed for which no matter how extended the 'dead stick' glide, or how close to the shore, the airplane is assured to never reach land.

BIENNIAL FLIGHT REVIEW — A general review of flight regulations and a checkride review of airwork with a Certified Flight Instructor, who signs the pilot's log book and attests to the fact that the pilot is competent to go forth for another twenty four months and terrorize the friendly skies of United. Sunday Pilots look forward to this test, since some of the material may be new to them, having been absent because of illness the day their instructor was going to cover a particular subject.

BIG SIX — The 'Big Six' refers to the six basic flight instruments used during instrument attitude flying. They are arranged in a basic 'T' configuration, and by scanning them, the airplane can be flown without visual reference to the outside. They consist of the following indicators: airspeed, attitude, direction, altitude, vertical speed, and turn coordinator. These are the instruments that Sunday Pilots concentrate on watching, when the outside view through the windshield gets either too scary or too dark.

CARBURETOR HEAT — Carburetor heat is warm air diverted from the exhaust system of an airplane engine into an enclosure surrounding the carburetor. It is used during icing conditions to keep the butterfly valve of the throttle (in the throat of the carburetor) from forming ice, thus choking incoming air, which would 'kill' the engine. The Carburetor Heat knob is the knob that Sunday Pilots mistakenly pull to the 'ON' position, for cabin heat, during flight on a cold day.

CESSNA 170 — A vintage four place, high wing, single engine, all metal, monoplane of monocoque construction, with conventional landing gear configuration (tail wheel). Built during the sixties and seventies, this airplane is still used in Alaska by bush pilots. It was the model that became the Cessna 172, when a nose gear replaced the tail wheel, making it possible for Sunday Pilots to land without crashing.

CESSNA 172 — Built since the sixties, the Cessna 172, named the Skyhawk, was the first model to take a 'name' as well as a numerical model designation. It is a four place, high wing, single engine, all metal, monoplane of monocoque construction. Because of its nose landing gear, the C-172 was the first general aviation airplane considered 'modern.' It is now the most popular family airplane in the world, not only out selling all other makes and models, but, statistically, the safest to fly, as well. Its range is four and a half hours and it cruises at one hundred and thirty five miles per hour.

CESSNA 182 — Built since the seventies, the Cessna Skylane took the popular Skyhawk and put a larger engine in it for a thirty mile per hour increase in speed. The 182 is the perfect family single engined airplane. It's gear is fixed (not retractable) making insurance rates just half as expensive as other high performance airplanes.

CESSNA 210 — Built since the fifties, the Cessna 210 (later called the Centurion) is the fastest six place, all metal, high wing, single engine, airplane that Cessna manufactured. It has a 285 horse engine and retractable gear, and a range of over one thousand nautical miles. This airplane is certainly beyond the price range of most Sunday Pilots. It's flown by doctors and lawyers and those lucky enough to have won the Lottery, and can afford the rental fees.

CFI — Certified Flight Instructor. This certification allows a person to instruct perspective pilots in basic or primary flight. These are usually people in their twenties who are trying to log as much 'pilot in command' time as they can so that in a few years they can qualify for a high paying airline job. Most don't make it, ending up in non-flying jobs for the remainder of their lives (to pay their rent). But, sometime, if they miss flying too much, they become Sunday Pilots like the rest of us 'working stiffs.'

CFII — Certified Flight Instructor Instruments. This certification allows a CFI to instruct licensed pilots in instrument flight, so that the private pilot may qualify for his instrument rating enabling him to fly fearlessly into bad weather and terminating the flight into 'cumulo granite' (a cloud with a mountain inside it.)

CHEROKEE 140 — The Cherokee 140 was Piper Aircraft's answer to Cessna's 'modern' Skyhawk. It is a four place, all metal, low wing, fixed gear, monoplane of monocoque construction. It's range is four hours at a speed of one hundred and ten miles per hour. Piper always named their airplane models beginning with the 'Cub' through the 'Vagabond,' 'Clipper,' Pacer,' and 'TriPacer.' The Cherokee series of models began after the TriPacer of the fifties and continued into the eighties. The numerical designations of the Cherokee series refers to the model which directly refers to that model's engine horsepower; thus, the PA-235 had a horse power rating of 235 hp.

CROSSWIND LEG — The Crosswind leg of an Airport Traffic Pattern. This leg is at right angles to the active runway at its departure end. As soon as four hundred feet of altitude is achieved straight ahead on takeoff, the airplane is turned ninety degrees to the left or right from the runway alignment; thus, putting the airplane in a 'crosswind' situation in regards to the wind direction. Don't worry if it's hard to visualize, Sunday Pilots have the same problem, even when they're supposed to know which way to go. (See Base Leg.)

DEAD RECKONING — Dead reckoning is the navigation of an airplane solely by means of computations based on airspeed, course, heading, wind direction and speed, and elapsed time. Sunday Pilots have learned to use redundant means to navigate in case the 6th Grade arithmetic of their dead reckoning turns out wrong or doesn't turn out. Several alternate means of navigation are following state highways, follow railroads, or following slow flying airliners; hoping that they'll land somewhere close to the Sunday Pilot's destination.

DF — Direction Finder. A radio receiver equipped with a directional sensing antenna used to take bearings on a radio signal. Some of these radios are airborne (see ADF), and others are used on the ground by Flight Service to obtain a 'fix' on a Sunday Pilot requesting orientation assistance (meaning he's lost, again! and can't find the airport).

DIRECTIONAL GYRO — A gyroscopic navigation instrument sometimes referred to as a 'gyro compass.' It replaced the 'whiskey' or wet compass as the primary instrument for directional flight during instrument flight conditions, because the wet compass moves erratically during turns, taking a long time to settle down on a heading; even longer than it takes a Sunday Pilot to settle himself down on turns.

DME — Distance Measuring Equipment. A 'black box' used in the airplane to measure the distance from the airplane to a navigational aid on the ground. Before Global Positioning System equipment, DME was the only way a pilot had of knowing in real-time, his ground speed and 'estimated time of arrival' to the next fix, without using dead reckoning or LORAN C.

DOWNWIND LEG — The Downwind leg of an airport traffic pattern is a flight path parallel to the landing runway, in the same direction as the wind, or in the opposite direction to that of landing. The Downwind leg normally extends between the crosswind leg and the base leg. It's a good thing that the wind knows what direction it's blowing and what it's descriptor is named, because Sunday Pilots still have a hard time remembering all this technical stuff while trying to land their airplane without crashing. (Also see Crosswind Leg, Base Leg, Upwind Leg, and Final Approach.)

EAA — Experimental Aircraft Association. The EAA is a large association of amateur airplane builders, referred to as 'home builders,' of experimental airplanes. The association is comprised of builders, pilots, and aviation enthusiasts.. It is headquartered in Oshkosh, Wisconsin on the grounds of Wittman Field.

ELT — Emergency Locator Transmitter. A radio transmitter attached to an airplane's structure that operates on its own power source and automatically transmits a signal on an emergency frequency, if the airplane crashes. The signal can be received by search and rescue airplanes as well as the overhead search and rescue satellite system. Sunday Pilots hope that their ELT's are in good working order, in case there's no ASR or Approach Control or Flight Service Station to locate their position, when they're lost and low on fuel. At least if they have to land somewhere other than at an alternate airport, and are able to survive the crash, with an ELT aboard, they hope that they may eventually be found and rescued.

ERCOUPE — A vintage two place, side by side, single engine, metal and fabric, low wing, monoplane of monocoque construction first built in the middle thirties, and after WWII, continuously built until the late sixties. The airplane was thirty years ahead of its time with advanced technology that was only reintroduced to aviation in the sixties. The airplane has a range of about 450 miles and a speed of one hundred and five miles per hour. It was known in the thirties as the world's safest airplane, because its design wouldn't allow it to stall or spin.

FAA — Federal Aviation Administration. The government agency that controls the rules, regulations, and certification of pilots, mechanics, and airplanes. It is the one agency of U.S. government that says 'it's here to help the pilot.' That phrase usually motivates pilots to run for the door and keep running. Over the years, because of the FAA's 'help' general aviation has seen the drastic decline of their enterprise. Sunday Pilots feel about the FAA, the way Russians use to feel about their Czar: 'We love him, as long as he stays far, far, away.'

FAA EXAMINER DESIGNEE — A designated flight examiner appointed by the FAA to certify general aviation applicants for various licenses and ratings.

FBO — Fixed Base Operator. The operator of a flight school and airplane rental. One stop shopping for a person that wants to learn to fly. The FBO offers on staff, flight instructors, airplanes for rent to student pilots with instructors aboard, or to licensed pilots, as well as maintenance departments to keep the rental airplanes in good repair. As a student pilot, the FBO is where he brings eighteen wheelbarrows full of money to trade for flying lessons.

FERRY FLIGHT — A flight for the purpose of returning an airplane to its home base; delivering an airplane from one location to another; or moving an airplane to or from a maintenance facility. Not to be confused with the flight of a 'gay' pilot.

FINAL APPROACH LEG — The final approach leg of an airport traffic pattern is a flight path in the direction of landing along the extended runway centerline. The final approach normally extends from the base leg to the runway touchdown. This leg is a Sunday Pilot's favorite, since it means that he'll soon be on the ground where it's safe and that 'happy hour' can commence.

FLAG — A warning device, usually in the shape of a small red rectangle, incorporated into certain airborne navigation and flight instruments, indicating that the instruments are inoperative or otherwise not functioning properly; or that the received signal operating the instrument is of such marginal quality that it falls below acceptable values. For the Sunday Pilot the 'flags' usually come out when the pilot is lost and low on fuel, over unfamiliar territory, and hundreds of miles from civilization.

FLIGHT PLAN — A documented plan of an intended route of flight normally filed with Flight Service. A VFR flight plan can be filed twenty three hours and fifty nine minutes in advance of the flight. IFR flight plans can be filed as early as VFR plans, or as late as just a half hour prior to the flight. The plan identifies the airplane by tail number, the navigation equipment onboard, the cruise speed, hours of fuel onboard, the route of flight, how many people onboard, and a description of the airplane's markings. Sunday Pilots either forget to file flight plans or file them, but forget to activate them, once the flight commences, thus ensuring that if they crash they will never be found.

FLIGHT SERVICE STATION (FSS) — The flight service station, sometimes referred to as 'Flight Service,' is not a gas station for airplanes, it is an FAA facility that provides preflight weather briefings either in person or over the telephone. FSS provide inflight weather updates, enroute communications, VFR search and rescue service, assists in lost aircraft and aircraft in emergeny situations, relays air traffic control clearances, originates notices to airmen, broadcasts aviation weather, receives and processes IFR Flight Plans and monitors navigational aids.

GRAVEYARD SPIRAL — An inadvertent flight maneuver caused by releasing the control wheel of the airplane and leaving it unattended for more than a few seconds (to open maps or look up radio frequencies or other chores). An airplane may not be perfectly balanced; it should have more fuel in one wing tank than another, or the pilot may be flying alone. When the control wheel is released and the airplane is allowed to seek its center of balance, the airplane naturally falls off into a shallow turn to the left which. If left unattended for too long, it steepens to the point of ending in a high speed vertical dive that is unrecoverable.

GROUND CONTROL — At a controlled airport with an operating control tower, no airplane is allowed to move on the surface of the field without prior permission. Ground Control is a control tower position that controls, by radio, all ground movements within the taxiways and flight ramp areas, where airplanes normally travel; including controlling the airport operations ground crew mobile trucks. Ground Control does not control the airplane movements on the active runway.

HSI Horizontal Situation Indicator — This instrument portrays in a plan view four instruments in one: VOR, Directional Gyro, localizer, and glide slope. Not only does is reduce the amount of instrument space on the panel, but is also very convenient to scan only one instrument to get both the heading and course, as well as the localizer and glide slope during a precision approach. The airplane representation in the middle of all this helps the pilot visualize the airplane's relative situation over the earth. Sunday pilots usually can't afford an instrument that costs five or six thousand dollars, no matter how much panel space it saves. And beside, when it breaks down, it can't be cheap to fix with all the moving parts.

IFR INTERSECTION — A point in space, defined by any combination of courses, VOR Radials, or bearings of two or more navigation aids. Each intersection is named and is published on various charts and approach plates.

ILS — Instrument Landing System. A precision instrument approach system that normally consists of the following electronic components and visual aids:
1) **Localizer**, the component of the ILS that provides course guidance to the runway;
2) **Glideslope**, the component that provides vertical guidance, based on components emitting an electronic signal that provides a three degree approach angle;
3) **Outer Marker Beacon**, an electronic nondirectional beacon, located at or near the glideslope intercept altitude of an ILS approach, normally four to seven miles from the approach end of the runway;
4) **Middle Marker Beacon**, an electronic non directional beacon that defines a point along the glideslope of an ILS, normally located at or near the point of Decision Altitude;
5) **Approach Lights**, an airport lighting system that provides visual guidance to landing airplanes by radiating light beams in a directional pattern by which the pilot aligns the airplane with the extended centerline of the runway on his final approach for landing.

LAKE AMPHIBIAN L-2 — The Lake Amphibian or flying boat is a four place, single hulled, all metal, single engine, midwing flying boat, manufactured by the Lake Airplane Company. The single, pusher' engine sits on a four foot-high pylon, above and behind the cabin. It cruises at 120 miles per hour with the two hundred horse engine; and has a four hour fuel capacity

LORAN C — Long Range Radio Navigation. An electronic navigation system that provides an airplane, in flight, with its position from a selected navigational aid or geographic point on the surface of the earth. It is accurate to within sixty feet. The Sunday Pilot uses it mainly to find the correct heading, ground speed, and estimated time of arrival to his destination airport.

LOCALIZER — The electronic beam broadcast by part of the equipment of an Instrument Landing System (ILS), that aligns itself to the centerline of the runway, so that during instrument conditions, by intercepting and centering the 'localizer needle' in the cockpit indicator, the airplane can 'home on' and fly to the runway with perfect alignment.

MAGNETIC COURSE — The proposed flight path direction relative to magnetic north. Victor Airways are depicted on aeronautical charts as magnetic courses; those Airways are generated by VOR stations, as discrete electronic beams aligned from magnetic north.

MANIFOLD PRESSURE GAUGE — An engine cockpit instrument that measures engine intake manifold pressure. The gauge is used as an indicator to adjust throttle settings on high performance general aviation airplanes with a constant speed propeller. The gauge measures the amount of air, in inches of mercury, entering the engine for a particular power setting.

MOA — Military Operational Areas. Areas depicted on navigation charts to be avoided by general aviation airplanes. Military pilots use these areas to practice gunnery, bombing, and combat tactics. The Sunday pilot usually finds himself cutting across one of these areas when they are 'hot.' Needless to say, flying through the middle of a couple of jet fighters while they're 'dog fighting,' is about as life threatening as running into the jungle clearing of a Mau Mau camp meeting, and spitting on the Witch Doctor.

MOONEY CADET — A two place, side by side, all metal, low wing, single engine monoplane of monocoque construction. The Cadet is a derivative of the Ercoupe. It had a conventional single vertical tail and an upgraded canopy and interior. Mooney Aircraft bought the rights to the Ercoupe in the seventies and modified the former Ercoupe to be the new Mooney trainer airplane. The Cadet has a speed of one hundred and ten miles per hour and a range of four hours. It also doesn't fit any Sunday Pilots that are larger than a twelve year old child.

NAVIGATIONAL AID — Any visual or electronic device airborne or on the surface that provides point-to-point guidance information or position data to airplanes in flight. To a Sunday pilot navigational aids also include moving trains, large signs painted on the sides of hangars, buildings, or water towers, telling you what airport or city you just landed at.

NON DIRECTIONAL BEACON — A low powered radio transmitter that emits its signals in all directions, so that airplanes can 'home on' its location with their radio compass indicator. It is an anachronistic artifact from the golden age of aviation of the thirties and forties that was then used for navigation purposes.

NOTAM (Notice To Airmen) — This is an advisory from Flight Service to pilots to be used during the planning phase of a flight that descibes the local conditions that exist at an airport; as to the status of equipment, taxiways, and radio frequency changes, that will have a short term effect on the flight. Sunday pilots use NOTAMs to see if the airport restaurant or bar is open or closed.

NO GYRO APPROACH — A radar approach and or vector provided to the pilot in case of a malfunctioning gyroscopically controlled direction indicators. Instead of providing the pilot with headings to be flown, the radar controller observes the airplane's ground track by radar, and issues control instructions: 'turn right, turn left, stop turn,' as appropriate.

PIPER PACER — The Piper Aircraft model PA-20, or Pacer, is a four place, single engine, all fabric, high wing, monoplane, with conventional landing gear (tail wheel). It was the last tail wheeled airplane model that Piper built. It had a range of four hours and a cruise speed of one hundred miles per hour. It was manufactured from 1950 through 1957. It is still used in Alaska as a float plane in the bush.

PIREPS (Pilot Reports) — These reports are issued by Flight Service and document current flight conditions over discrete routes or areas. A pilot flying in a certain area or to a certain area can call Flight Service on the radio and ask about pilot reports for his route of flight. If any current reports have been received, the pilot can be apprised of what 'real world' weather conditions have been encountered by pilots who, only minutes before, flew over the same route.

PIC (Pilot In Command) — PIC 'time' is the time that a pilot is sole manipulater of the airplane's controls. For CFIs and CFIIs instructing students, all of the student's flying time can be recorded as 'PIC time' for the instructor. Thus, flight instructing is an easy way to collect PIC time for that always potential airline job. The airlines require a certain minimum PIC time as proof of flight experience.

PITTS S-2A — A classic single engine, two place, all fabric, aerobatic midget biplane, capable of world class competition, and designed by Curtiss Pitts in the late fifties. It is still in production today, both factory-built or by amateur homebuilders using plans. Originally, the Pitts was a single seater, but the S-2A allows an instructor to fly with a student to teach aerobatics. Various engines have been fitted for both the single seater and the two seater; and vary from 150 to 1,000 horse power. The airplane is capable of climbing vertically, from ground level, to more than three thousand feet, before its forward speed pays off.

PILOTAGE — This is the way pilots during the twenties and thirties navigated; by looking down at landmarks on the ground, and comparing what they saw there to what similar landmarks were symbolized on the map in their laps. When a landmark on the ground agrees with a landmark on the map, the airplane's position is authenticated. Just how accurate this kind of navigation can be is documented in Chapter 2, when, as a student pilot, I used this method to locate my position while flying over the Arizona desert to Phoenix.

POSITIVE CONTROL AREA — In the National Airspace System there are seven classes of airspace: Class A through Class G. Several classes are areas of positive control, where a pilot may not enter without prior permission, and once admitted, is under positive radar control by the cognizant authority.

PROPELLER (PITCH) LEVER — A lever that controls the

constant speed propeller's variable pitch. Adjusting the lever is like shifting gears in a car; it sets the speed of the propeller at a certain rpm, and makes the propeller constantly change its pitch to adjust to maintain the desired 'constant speed' rpm of the engine, during all attitudes of flight, thus making the engine more efficient.

RADAR (Radio Direction And Range) — Radar is the electronic method of gathering information about distant objects by means of reflected radio waves. When electronically processed, the radio waves can be used to determine the distance and bearing of an object from the radar transmission site.

ROTATING BEACON — A visual navigation aid operated at many airports. It is a rotating light, usually found on the top of the Control Tower. The light has a green lens on one side and a white lens on the other, thereby alternating or flashing green and white as it rotates. It is aimed at an elevation of three degrees above the horizon, so that low flying aircraft can use it to visually locate an airport.

SPECIAL VFR — A VFR clearance, issued by the tower, upon request, when the field is below VFR minimums. The horizontal visibility can be limited to one mile; but there can be no ceiling, and the skies must be clear of clouds. It generally allows one airplane at a time permission to either take off and climb out to VFR conditions, or to approach and land in conditions less than VFR.

STALL SPEED — The air speed below which the airplane cannot sustain flight and falls out of the sky. The physics of flight effects all airplanes in the same way. Size does not matter. Whether an airplane is a Boeing 747 or Piper Cub, when their speed falls below a certain rate, the amount of air passing over and under the airfoil of the wings is not sufficient to sustain flight, and the airplane takes on the aerodynamic characteristics of a Blacksmith's anvil.

SECTOR CONTOLLER — An air traffic controller responsible for controlling airplanes through a specific geographic 'sector' of the National Airspace System. Normally, a Sector Controller works in an Air Route Traffic Control Center, usually referred to as 'Center,' controlling airplanes on Instrument Flight Plans. This is also the person who, if workload permits, could offer Sunday Pilots radar 'flight following,' that keeps IFR traffffic from running into Sunday Pilots who are flying VFR only.

SIGMET — Significant Meteorological Information. An inflight weather advisory that amends an area forecast concerning weather that is dangerous not only to Sunday Pilots, but pilots flying the largest airplanes in the world. In other words, Sigmet condition are those that are harmful to everybody in the sky — conditions such as thunderstorms, tornadoes, typhoons and hurricanes.

SOFT FIELD LANDING — A landing maneuver used to land on a soft surface, such as snow, mud, tall grass, sand, tundra, or gravel. The object of this type of landing is to support the weight of the airplane with the wings for as long as possible during the landing roll, and to delay the weight transfer to the wheels during the rollout, until the airplane attains the slowest possible speed.

SPAM CAN — A slang term that refers to any of the 'production' or factory built metal airplanes and implies that operating them as a pilot, entails a certain commonality of boredom and unexcitement. The implication is that all factory built production airplanes are pretty much the same and terribly unexciting to fly; as opposed to flying tail draggers, that require an extra level of skill to land without ground looping (crashing).

SQUAWK — To 'squawk' a descrete code, is to set a four digit code number sequence into an airplane's radar transponder. ATC issues discrete codes to each airplane they control in their sector. This code then generates a discrete shape or image on the receiving radar screen,

STANDARD DAY — Standard Day refers to standard conditions that are used as a baseline for all airplane performance charts. Its baseline is: a pressure altitude of 29.92 inches of mercury at sea level; and a temperature of 59 degrees F, or 15 degrees Celsius.

TAIL DRAGGER — An airplane whose landing gear configuration is a main landing gear under each wing and a tail landing gear at the tail end of the fuselage. This configuration is less stable than that of modern tricycle gear airplanes, since the main gear of the tail wheeled airplane is behind the center of lift, thus allowing the airplane to 'weather vane' upon landing during a crosswind condition. To be able to control a tail wheeled airplane, extra flight instruction is required of its pilot, as the airplane is unstable during the landing and taxi phase of flight.

TAXI — The movement of an airplane under its own power on the surface of the airport.

TERMINAL CONTROL AREA — An area of positive air

traffic control, normally surrounding a high density commercial airport. This area of positive radar control is needed to keep airliners and general aviation airplanes separated during approach and landing phases of flight. This area can't be penetrated without prior permission by radio from the controlling agency.

TIE DOWN — A parking space for general aviation airplanes. It consists of a tie down cable strongly fastened to the parking ramp, with either chains or ropes attached to the cables, so that each wing can be 'tied down' as well as the tail end of the fuselage. Once the airplane is tied down, it would take a windstorm of hurricane force to break the airplane loose.

TRANSPONDER — The radar transponder is the airborne portion of the air traffic control beacon (secondary radar) system. It is an airborne 'black box' transceiver located on an airplane's instrument panel, that responds to a coded interrogation signal from ground equipment, and causes the airborne transponder to reply automatically, sending back a coded signal. This causes the airplane to appear as a distinctive pattern on the radar controller's scope. Thus, a single engine general aviation airplane's symbol appears as large on the radar screen as that of a 747 airliner's symbol.

TRIM TAB — A section of the airfoil of an airplane's elevator. The trim tab operates in opposition to the direction of the elevator deflection and this opposing of the aerodynamic forces on the airfoil negates the pressure required by a pilot to maintain on the airplanes' control wheel. The trim tab is controlled by a small circular control wheel mounted either in the instrument panel or between the two front seats near the floor of general aviation airplanes. In the Pacer and TriPacer it is a crank located in the ceiling of the cockpit.

UNICOM FREQUENCY — A common frequency used by an FBO at an uncontrolled airport to issue airport advisories to airborne traffic in the vicinity, as well as pilots communicating with other airplanes in the airport traffic pattern. Each uncontrolled airport has a published discrete UNICOM radio frequency. At controlled fields, pilots use Unicom to communicate with FBOs when requesting services from them, such as line service for fuel.

UPWIND LEG — The Upwind Leg of an airplane taking off, is that direction against the wind that the runway's aligned with. It is in the direction of takeoff. Since this is the easiest airport traffic pattern leg to remember, it is seldom described or used in commands by the control tower. But, since it is the simplest to remember, it is probably the only pattern leg Sunday Pilots can recite by rote.

VASI — Visual Alignment Slope Indicator. This two-light indicator device, located on one side of the approach end of an airport runway, indicates when a landing pilot is either on the desired three degree glide slope to the runway, or is above or below it. It is comprised of a series of two lights, separated by a certain distance; each capable of either turning red or white depending upon the angle from which they are viewed.

When the pilot is on the glide slope, the nearest light is white and the farthest is red. When the airplane is above the glide slope, both the nearest and farthest lights are white. When the airplane is too low, both the nearest and farthest lights are red. For Sunday Pilots, if the both lights are green, then they are already too low, as the light is being filtered through the grass infield.

VFR — Visual Flight Rules. The rules of minimum weather conditions that pilots use to determine whether it is safe to fly by visual cues, rather than on instruments. The minimum conditions for VFR flight are: a horizontal flight visibility of three statute miles, and a ceiling of at least one thousand feet above the ground. When either the ceiling or the visibility falls below those distances, the conditions are then referred to as 'below minimums,' and IFR or Instrument Flight Rules should be used.

VFR ON TOP — A condition where an airplane is on top of a continuous deck of clouds or undercast, above which, the weather is clear. This is a phase of flight that is normally an option for pilots flying IFR on an instrument flight plan. It is not only dangerous but illegal for a 'VFR only' pilot to be caught 'On Top,' as without a visible horizon or a view of the ground, a non-instrument rated pilot can easily become disoriented.

VICTOR AIRWAYS — The low altitude (non jet) air routes of flight depicted by connecting VOR stations via magnetically directed electronic beams, from one station to the next. These route segments are depicted on navigational charts by numerical designators, like the numbers of the interstate highway system on road maps.

VOR — Very High Frequency Omnidirectional Range. The federal airway system is based primarily on the system of VOR stations throughout the country. The extensive system consists of several hundred ground stations that transmit navigational course guidance that are used by airplanes in flight.

WATER RUDDER — On flying boats or float planes, the water rudders are steerable rudders attached to the ends of a float plane's pontoons or floats; or to the rear of the hull of a flying boat. The rudders, like those on boats, steer the airplane while taxiing on the water. The rudders have a cable mechanism that retracts them when the airplane is ready for flight.

WING OVER — A flight maneuver for quickly reversing direction in an emergency situation, like flying into a box canyon. It consists of a steep pull-up, to a lazy split 'S' and a half roll to level, reminiscent of trying to do a sloppy 'Immelmann' with insufficient entry speed. It's the kind of maneuver that Sunday Pilots were born to fly.